Advanced
Stereo System Equipment

Advanced
Stereo System Equipment

Lloyd Hardin

RESTON PUBLISHING COMPANY, INC.
A Prentice-Hall Company
Reston, Virginia

Library of Congress Cataloging in Publication Data

Advanced stereo system equipment.
Includes index.
1. Stereophonic sound systems. 2. Sound—Recording
and reproducing—Equipment and supplies. I. Title.
TK7881.8.M48   621.389'3   79-23426
ISBN 0-8359-0166-1

1   3   5   7   9   10   8   6   4   2

PRINTED IN THE UNITED STATES OF AMERICA

Contents

Preface

This work is both a textbook and a technical handbook; it will find ready acceptance among audio technicians and business managers in stereo shops and by sales personnel looking for understandable information that will make them effective in customer relations. Instructors in all technical schools that offer audio courses will quickly recognize that this text is a "quantum jump" ahead of its competitors.

Advanced stereo systems and conventional stereo systems use the same components and units, except that the equipment for advanced systems is designed to standards of higher performance.

As well, many advanced systems have sophisticated processors and devices that optimize sound reproduction. Because the acoustics of the listening area are an important part of an advanced stereo system these parameters also receive appropriate attention in the following pages. The first chapter in this text provides a general introduction to advanced stereo systems and discusses basic system options. Volume expansion and compression are explained, the double-Dolby system is illustrated, and the various kinds of multiple-head tape machines are described. The second chapter, which is about advanced stereo acoustics, covers sound mirrors, phantom areas, and virtual sound sources, and explains the significance of sonic standing waves and acoustic resonances. This chapter also describes "Sound traps" and the role of frequency equalizers in partial correction of acoustic anomalies.

Chapter 3 develops the nine classes and transfer modes of audio amplifiers and analyzes volume expansion and compression. It also discusses basic biamplifier design and its inherent advantages as well as the new IHF dynamic headroom characteristic. Chapter 4 discusses high-performance integrated receivers, basic design features for high-performance stereo-FM tuners, the capture effect, and the usefulness of an audioscope monitor in checking for multipath propagation. In addition, stereo separation displays on an audioscope screen are illustrated. Chapter 5 which reviews high technology tape machines, covers metal-particle tape and its requirements, the principles of digital recording and playback, system organization, and advantages and disadvantages of automatic level control. The sixth chapter provides a survey of specialized stereo equipment, including wireless remote control, color organs, and strobe lighting, along with the bucket-brigade audio delay system which is more fully analyzed. Basic characteristics of acoustic screens are noted and biphonic sound is technically compared to monophonic, stereophonic, and quadraphonic reproduction.

Advanced stereo tests and measurements are explained in the seventh chapter which gives procedures for checking an acoustic environment with a sound-level meter and associated instruments, explains determination of bass, midrange, and treble acoustic profiles for a listening area, and describes acoustic resonant-frequency measurements. Distinctions are made among acoustic, electronic, and psychophysical stereo separation, and the basis of psychophysical separation is detailed. The more conventional stereo tests and measurement are also included. Chapter 8 surveys sophisticated recording techniques. It covers the following topics: advantages of balanced microphone lines; balun requirements for system compatibility; control of microphone proximity effects; advanced stereo recording techniques with the dual stereo microphone (coincident microphone); microphone placement for various instrumental and vocal groups; and the use of acoustic screens for optimizing recording characteristics. Notes are included on good practices in recording of radio programs.

Chapter 9, public-address sound equipment is discussed with respect to general requirements in various acoustic environments and the practical aspects of speech-reinforcement techniques. Principles of controlled time delay for optimization of articulation are explained, and specialized speaker requirements are detailed with notes on system equalization. Problems of acoustic feedback are outlined with respect to microphone acceptance patterns and to system frequency response. Practical mixing techniques are included.

The tenth chapter covers miscellaneous audio equipment: intercommunication systems, wireless microphones, telephone answering equipment, telephone dialers and speech amplifiers, handset pickup units, electronic megaphones, surveillance "bugs," specialized microphones, audio polygraphs, audiometers, electronic stethoscopes, audio analysis equipment such as the sound spectrograph, and speech synthesizers. The application of high-frequency sound waves to supplement the use of X-rays in medical diagnosis of internal diseases and abnormalities is also noted.

As the preceding paragraphs indicate, this volume offers considerable information on audio and stereo equipment. It is therefore an invaluable reference book for all serious stereophiles, from beginners to experts.

Advanced
Stereo System Equipment

1

Introduction to
Advanced Stereo Systems

BASIC ARRANGEMENTS

An advanced stereo system consists of all the components employed in a conventional stereo system, plus processors and devices that optimize sound reproduction. Most advanced stereo equipment is more sophisticated and costly than conventional equipment, but a few items may be used in either system. Auxiliary processors and devices are added to advanced stereo systems to improve signal characteristics, to accommodate the acoustics of the listening area, and to provide increased operating flexibility and convenience. Therefore, it is advisable to select speakers during the first stage of planning an advanced stereo system; speaker performance factors are outlined in another section. Next, choose an amplifier that provides adequate power output to drive the speakers to their maximum rating. An integrated receiver contains an FM/AM tuner, preamplifier and power amplifier in the same cabinet. The receiver is generally used with an external turntable (record player), external tape player, and optional earphones.

Component systems are preferred by some advanced stereo aficionados. For example, a chosen set of speakers may be grouped with a suitable power amplifier, a preamplifier, a turntable, a reel-to-reel tape deck and/or an eight-track deck, or a cassette deck. Frequency equalizers are generally included in advanced stereo systems; and audio processors and ambience time-delay units are becoming more common. High-fidelity sound reproduction requires a frequency response that is uniform within ± 1 decibel (dB) from 20 hertz (Hz) to at least 20 kilohertz (kHz), with a harmonic distortion value of less than 1 percent at any frequency within this range. A block diagram of an

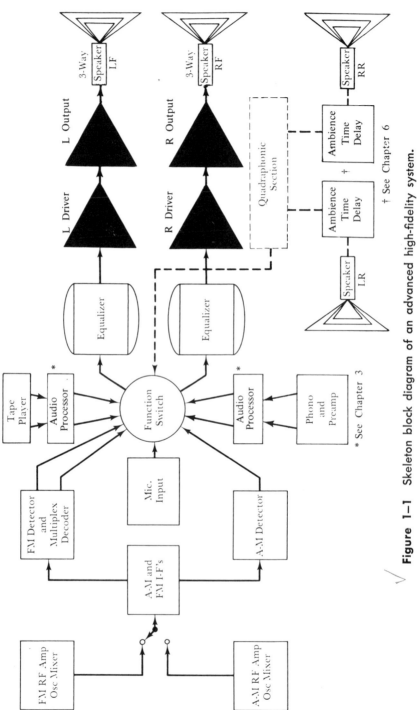

Figure 1–1 Skeleton block diagram of an advanced high-fidelity system.

advanced high-fidelity system appears in Figure 1–1. Other advanced stereo enthusiasts prefer an all-in-one console system. This unitized design features a single large cabinet containing a pair of speakers and all the system electronics. A typical console system has a turntable, FM/AM tuner, stereo decoder, preamplifier, power amplifier, two speakers, and various accessories. Evaluation factors for advanced stereo systems are noted in Chart 1–1.

√ CHART 1–1

EVALUATION FACTORS FOR ADVANCED STEREO SYSTEMS

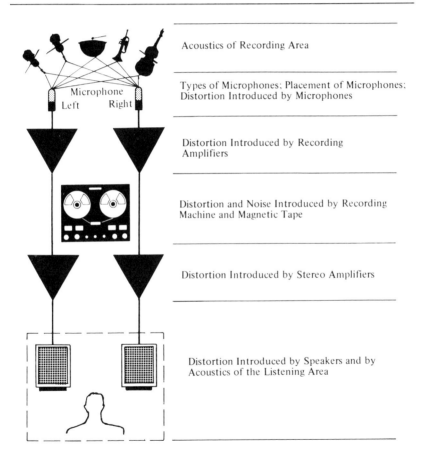

Acoustics of Recording Area

Types of Microphones; Placement of Microphones; Distortion Introduced by Microphones

Distortion Introduced by Recording Amplifiers

Distortion and Noise Introduced by Recording Machine and Magnetic Tape

Distortion Introduced by Stereo Amplifiers

Distortion Introduced by Speakers and by Acoustics of the Listening Area

Rack-mounted equipment, which permits the hi-fi buff to "tailor" a matched system in accordance with personal preferences, is becoming more widely used in advanced high-fidelity systems. The example in Figure 1–2 shows a tuner, cassette deck, turntable, preamplifier, and

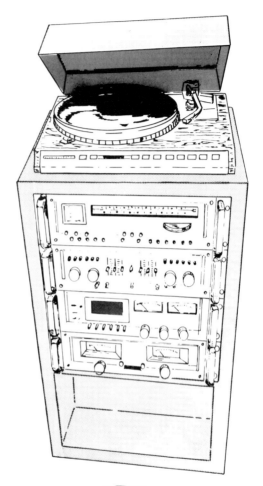

Figure 1-2 Typical rack-mounted high-fidelity system.

power amplifier mounted in the rack. Another versatile system with matching amplifiers and optional plug-in modules that permits the audiophile to "custom design" an advanced stereo system is called the modulus system. Basic system options are briefly described in Chart 1-2. The compact or modular system is in wide use. The principal components of a compact system may be an FM/AM tuner plus a turntable; another design may include a record changer mounted on top of the main unit and protected by a clear plastic cover. A compact or modular system is a comparatively economical design.

√√ CHART 1–2

BASIC SYSTEM OPTIONS

All-in-one System

Console (unitized) design, wherein a single large cabinet contains a pair of speakers and all of the system electronics. A typical unit comprises a record player, an FM/AM tuner, stereo decoder, preamplifier, power amplifier, and two speakers.

Compact (Modular) System

Consists of a turntable and stereo amplifier on the same base, with separate speaker enclosures. Some designs include an FM/AM tuner on the same base with the turntable and amplifier. A few include a tape deck, also. The most elaborate designs provide a 4-channel amplifier for reproduction of discrete 4-channel tapes.

Rack-mounted System

A flexible system that features removable components mounted in a vertical rack. A typical system comprises a tuner, cassette deck, turntable, preamplifier, and power amplifier. A record storage space is usually provided at the bottom of the rack.

Modulus System

A versatile system with matching amplifiers and optional plug-in modules that permits the audiophile to "custom design" a system to meet individual preferences. This system is easily changed or updated, as desired. For example, Dolby FM may be added to a basic assembly; a higher power amplifier may be employed; quadraphonic reproduction may be added by means of plug-in SQ or CD-4 modules. Other system variations are available.

Integrated Amplifier

An integrated amplifier is an audio-amplifier unit that contains both a preamplifier and a power amplifier. It often provides more control refinements and more inputs (signal input facilities) than an all-in-one system.

Integrated Receiver

An integrated receiver contains an FM/AM tuner, preamplifier, and power amplifier in the same cabinet. It is operated with an external turntable, external tape player, external speakers, and optional earphones.

Speakers

Small (bookshelf) types of enclosures have limited audio power capability, comparatively limited bass response, and approximately half the efficiency of large enclosures. Large reflex three-way or four-way enclosures have high power capability, extended low-frequency response, and moderate efficiency. Horn-type speakers have considerably higher efficiency; only folded-type corner-horn enclosures are sufficiently compact that they can be installed in residential listening areas.

Tape Machines

A tape recorder provides both recording and playback facilities; a tape deck lacks recording facilities. A tape deck does not have a built-in amplifier and must be used with an external amplifier and speaker system. Some tape decks provide recording facilities.

Record Players

Manual turntables lack automatic operation, but may provide somewhat higher fidelity than record changers. Most machines provide an anti-skate adjustment, speed adjustment, cueing lever, automatic shut-off, and a few include an automatic arm return.

SPEAKER CONSIDERATIONS

The interconnection diagram for an advanced integrated stereo receiver in Figure 1–3 shows two pairs of speaker terminals for connecting two Left (L) and Right (R) speaker systems (enclosures), and a third set of speaker terminals for connecting L and R electrostatic wide-range speakers. The impedance rating of this type of speaker often differs from that of a conventional dynamic speaker. All speaker lines are fused to avoid the hazard of overload damage. Basic speaker arrangements are depicted in Figure 1–4; two or more speakers (drivers) in an enclosure are called a speaker system. A two-way system has a woofer and a tweeter; a three-way system comprises a woofer, midrange speaker (squawker), and a tweeter. Extended bass response is commonly provided by a port; this design is called a reflex speaker system. A four-way speaker system (Figure 1–5) uses four speakers to cover the audio spectrum. If two woofers are included, it is termed a four-way/five-unit system.

Some enclosures contain a pair of midrange speakers, one of which is larger than the other. Normally, the size of a speaker is proportional to the amount of audio power that it can radiate. Speakers in an enclosure operate in association with crossover networks that direct suitable ranges of audio frequencies to each speaker. Level controls may be provided for the tweeter and/or midrange speaker to obtain correct tonal balance. Many speakers have both a minimum power rating and a maximum power rating. The minimum power rating denotes the lowest audio power level at which satisfactory performance can be realized in a small listening area. On the other hand, the maximum power rating denotes an audio power-input level above which

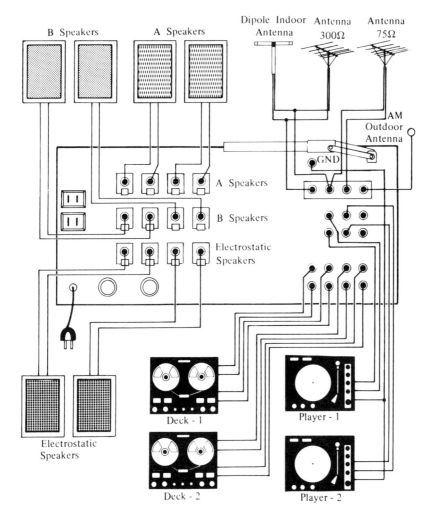

Figure 1–3   Interconnection diagram for the Lux Stereo FM/AM integrated receiver.

the speaker is very likely to be damaged. A speaker that is rated for a minimum power input of 6 watts may be rated for a maximum power level of 100 watts. The power rating of a speaker system is equal to the sum of the power capabilities of the individual speakers. For example, in a two-way speaker system, the woofer might be rated at 100 watts, the tweeter at 35 watts, and the speaker system at 135 watts.

When space is at a premium, as in automobile stereo systems, a single speaker having three-way coaxial construction may be used, as depicted in Figure 1–6. This design provides extended high-frequency

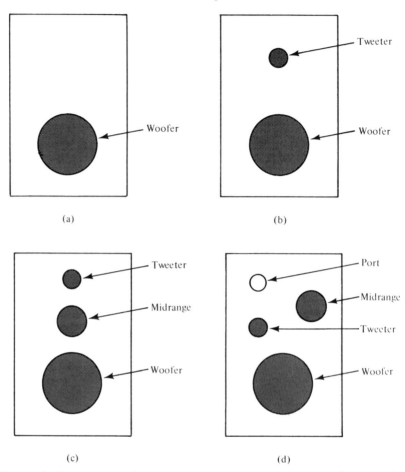

Figure 1–4 Basic speaker arrangements. **(a)** Enclosure with a single speaker; **(b)** two-way speaker system; **(c)** three-way speaker system; **(d)** reflex three-way speaker system.

response, although it lacks the uniformity of output over the entire audio-frequency range that is provided by a well-designed three-way speaker system. A large woofer may have a frequency response from 20 to 1500 Hz; a "squawker" has a typical frequency response from 800 Hz to 8 kHz; a tweeter covers approximately 4 to 16 kHz. Inductance-capacitance (LC) crossover networks are used in speaker systems to obtain smooth transition of output from the lowest to the highest audio frequencies. High-performance tweeters (Figure 1–7) provide extended high-frequency response up to 40 kHz. A piezoelectric (crystal) driver may be used, or a thin polymer diaphragm having an integral electrodynamic construction.

Figure 1–5 A four-way five-unit speaker system. *(Courtesy, Radio Shack)*

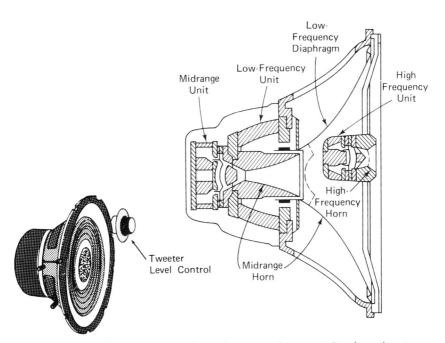

Figure 1–6 Three-way coaxial speaker is used in specialized applications, as in advanced auto-stereo systems.

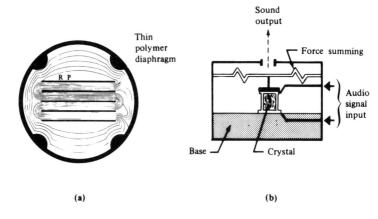

(a)   (b)

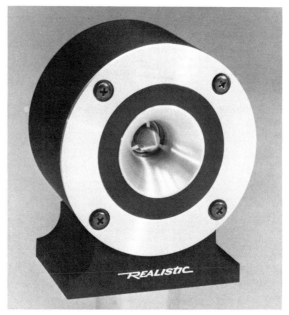

Figure 1–7 High-performance tweeter arrangements. **(a)** Polymer diaphragm/etched voice coil, 4–40 kHz; **(b)** piezoelectric design; **(c)** external appearance of a super tweeter. *(Courtesy, Radio Shack)*

The horn-type tweeter used extensively in speakers for advanced stereo systems is pictured in Figure 1–8, where a high-frequency driver energizes a short exponential horn. A representative speaker provides uniform output from 4 to 16 kHz, with a maximum audio power rating of 15 watts. The input impedance of the tweeter, like that of most woofers and tweeters, is 8 ohms. Note that a tweeter has a narrow "beam" of dispersion, compared with a cone-type woofer or midrange speaker. In other words, if the listener moves from directly in front of the horn to one side, he will experience a considerable reduction of sound intensity. However, this directional characteristic of a tweeter is modified substantially by the acoustics of typical listening areas. That is, a tweeter may seem to be radiating omnidirectionally because treble sound waves usually undergo extensive random reflections within a listening area. As detailed in subsequent paragraphs, a listening area occasionally requires acoustic treatment for optimum distribution of bass, midrange, and treble sound waves.

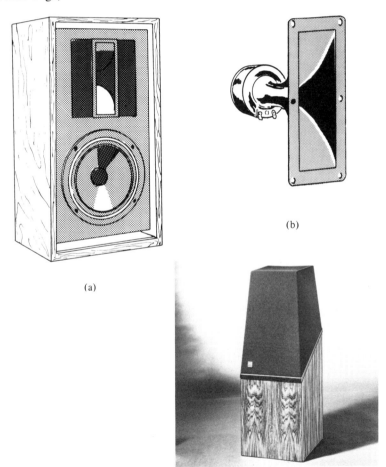

(b)

(a)

Figure 1–8 Horn-type tweeter. **(a)** Two-way speaker system with horn tweeter and cone-type woofer; **(b)** tweeter horn and driver unit; **(c)** appearance of a three-way linear-phase advanced-stereo speaker. *(Courtesy, Heath Co.)*

Obviously, balanced outputs must be provided by the woofer, squawker, and tweeter in a speaker system. Relative outputs depend upon the acoustic environment in which the speaker system is installed. Therefore, it is desirable to have relative level controls built into the system. Figure 1–9 shows a cone-type woofer with midrange and tweeter horn speakers. The multicell midrange horn develops comparatively wide dispersion from 800 Hz to 8 kHz, and the tweeter horn provides treble output from 8 to 25 kHz. The woofer provides extended bass output down to 20 Hz. Adjustable L pads permit the listener to vary the relative midrange and treble sound levels with respect to the bass level for a uniform sound field within a particular acoustic environment. It will be shown in a later section that relative levels of bass, midrange, and treble radiation are also a function of speaker location in a given acoustic environment.

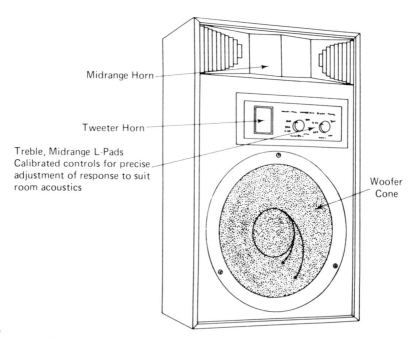

Figure 1–9 A three-way speaker system with adjustable treble and midrange L pads.

Speakers may have root-mean-square (rms) ratings—also known as sine-wave power ratings—or music (pulse) power ratings, or both. If a speaker power rating is stated in watts, without any qualifications, the rating refers to rms values of sine waves. This is a continuous or steady power rating; it represents the maximum power level that can be sustained indefinitely without damage to the speaker. On the other hand, a music-power rating refers to the peak-power value of a short pulse. This is a transient or discontinuous power rating; it represents the maximum peak-power level that can be withstood briefly without substantial distortion or damage to the speaker. A music-power rating pertains to sudden surges or attacks in musical waveforms. If a speaker were operated continuously at its music-power level, the voice coil would probably burn out. Most speakers have a rated input impedance of 8 ohms; some have impedance values as high as 16 ohms, and others have values as low as 4 ohms. To some extent, input impedance varies with frequency, and ratings are based on a speaker's impedance value at 1 kHz.

The cone in a large woofer may have a peak-to-peak excursion of ½ inch at maximum power output. This excursion corresponds to the limits of voice-coil travel, beyond which mechanical damage may occur. When more bass power output is required from a speaker system than can be provided by one woofer, a pair of woofers may be employed, as illustrated in Figure 1–10. In turn, each woofer provides one-half of the total acoustic power output, and the speaker system

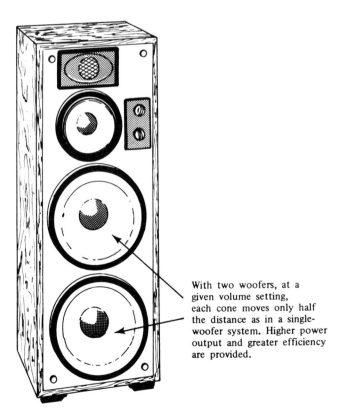

With two woofers, at a given volume setting, each cone moves only half the distance as in a single-woofer system. Higher power output and greater efficiency are provided.

Figure 1–10 Example of a three-way four-unit speaker system with treble and midrange level controls.

can transduce twice as much low-frequency audio power. This is an example of a three-way four-unit speaker system. As in the foregoing example, adjustable treble and midrange L pads are built into the system for contouring the treble and midrange output levels with respect to the bass output level; in this way the sound field in a particular listening area can be optimized.

An advanced stereo system may include an electronic organ like the one in Figure 1–11. In this case, the speaker system should be derated by 50 percent to avoid possible overload damage. This means that if an organ is rated for an output of 80 watts, the speaker system should be rated for a power input of 160 watts. This rule is based on the fact that the average power level in an organ output signal is often appreciably greater than the average power level in a tape, disc, or FM tuner output signal. Strong bass tones are developed by large organ pipe "voices," oboes, and so on. Loud bass reproduction requires comparatively large woofer cones and comparatively great displacement of the cones. A speaker system should also be derated when it has to reproduce sound from a substantially compressed orchestral recording. In other words, signal compression increases the average power of the waveform.

Figure 1–11   A top-performance electronic organ. *(Courtesy, Wurlitzer)*

Speaker output power requirements are related in a general way to the volume of a listening area. For example, if a listening area has a medium or average acoustic environment and a volume of 2000 to 3000³ feet, it can be served adequately by speakers having a power-input level of 15 watts rms per channel. Thus, the L and R speakers in this example would have a total audio power input level of 30 watts, maximum. For a room with a medium acoustic environment and a volume of 4000³ feet, speakers with an audio power rating of 20 watts rms per channel should be chosen. A very large listening area with a medium acoustic environment and a volume of 8000³ feet will require speakers rated for an audio power input of 45 watts per channel. If reserve power output is desired, speakers with higher power ratings must be employed. In a "lively" acoustic environment, the foregoing power requirements can be reduced by 50 percent. On the other hand, in a "dead" acoustic environment, the foregoing power requirements should be increased by 50 percent.

Ordinarily, stereo headphone (earphone) jacks, as seen in Figure 1–12, are available on stereo amplifiers and integrated receivers. Stereo headphones are illustrated in Figure 1–13. Some hi-fi connoisseurs prefer headphones because of their acoustical characteristics. If the cushions fit snugly, the listener will hear low bass tones that cannot be reproduced by speakers. Some audiophiles use headphones for privacy. Observe, too, that headphones provide the maximum possible separation of stereo L and R signals. Some hi-fi listeners do not like this high degree of separation, with its "hole-in-the-middle" acoustic effect. This effect can be overcome, however, by including a mixer pad in the headphone circuit, so that a chosen proportion of the L and R signals can be blended to develop a monophonic signal component.

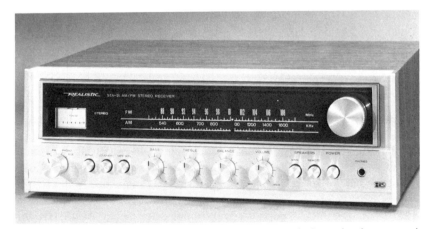

Figure 1–12 A jack for stereo earphones is provided on the front panel of this integrated receiver. (Courtesy, Radio Shack)

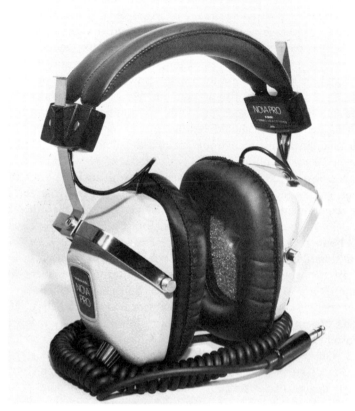

Figure 1–13 A pair of high-quality stereo earphones. *(Courtesy, Radio Shack)*

One form of mixer arrangement is depicted in Figure 1–14; basically, it is a balanced L pad that can be used either with an amplifier or with a pair of stereo headphones.

One primary technical distinction between a conventional stereo speaker and an advanced stereo speaker is that the advanced speaker has extended bass output. Although most conventional speakers have considerably attenuated bass output below 100 Hz, some have substantially attenuated bass output in the range below 150 Hz. At some lower frequency, any speaker's acoustic power output will drop 3 dB, or to one-half of its reference output level. Although this −3 dB frequency has complex technical aspects, here it will be termed the speaker's half-loudness point. A half-loudness frequency of 45 Hz, for example, qualifies a speaker for the advanced stereo category. However, other factors should be taken into account. Uneven frequency

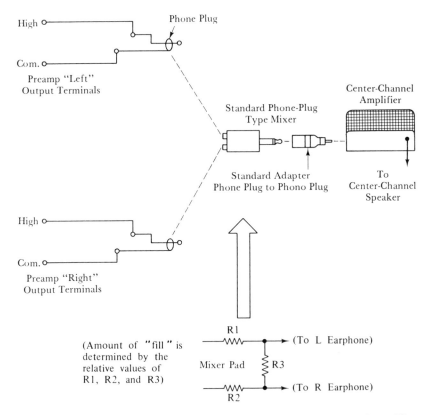

Figure 1–14 A balanced L pad is used to blend L and R signals to fill a "hole-in-the-middle."

response, appreciable harmonic distortion, or mediocre tone-burst response disqualify a speaker from advanced stereo classification. Note in passing that a speaker having somewhat deficient bass response, but high-performance characteristics otherwise, may reach advanced stereo level if it is located and oriented judiciously in the listening area. This topic is expanded in a following chapter.

AMPLIFIER CONSIDERATIONS

A basic amplifier channel comprises a preamplifier with input facilities for tuner, tape recorder, and turntable signals (Figure 1–15). More elaborate preamplifiers include input facilities for one or more microphones and for TV sound. Almost all preamplifiers provide equaliza-

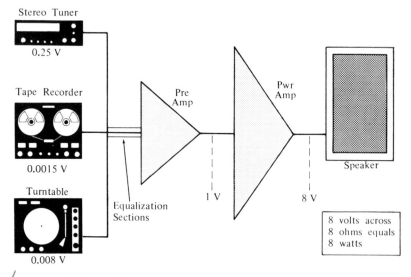

✓ **Figure 1–15**   Relative audio voltage levels in an amplifier system.

tion sections for tape player and turntable input channels. These frequency compensating circuits develop a bass boost characteristic that is opposite to the bass cut characteristic employed in the recording of tapes and discs. The preamplifier steps up the input signal level, such as 1.5 millivolts (mV), to a standardized output signal level, such as 1 volt. In turn, the power amplifier steps up this 1–V input signal into a higher output voltage such as 8 V. Note that if an 8–V signal is applied across an 8–ohm speaker, an audio power level of 8 watts is processed by the speaker. Since a three-way speaker may have an acoustic efficiency of 5 percent, it will radiate 0.4 watt of sound energy into the listening area.

As noted in Figure 1–16, a preamplifier has a typical input impedance of 50 kilohms (kΩ) and works into a representative load impedance (power amplifier input impedance) of 10 kΩ. Thus, the tape player in this example develops an input power level of 45 nanowatts (nW), and the preamplifier develops an output power level of 10 milliwatts (mW). In turn, the power amplifier in this example steps up this input power level of 10 mW to 8 watts. Note that although the power amplifier works into a load impedance of 8 ohms, the output impedance of the power amplifier is only a fraction of the load-impedance value. A low output impedance is required to adequately damp the speaker response and thereby to obtain good transient (toneburst) characteristics. As exemplified in Figure 1–17, various operating

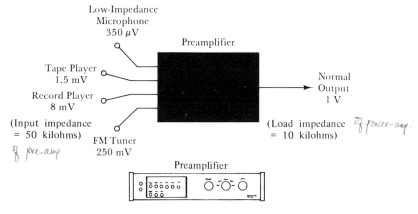

√√

Figure 1–16 Input and output voltages and impedances for a typical preamplifier.

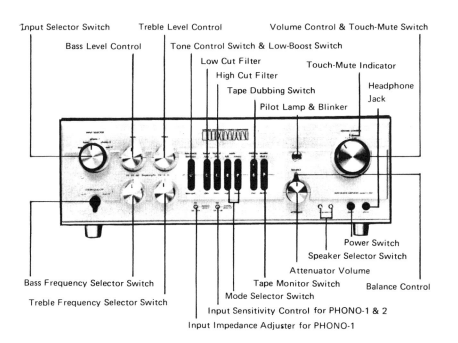

Figure 1–17 Example of amplifier front-panel controls. *(Courtesy, Lux Audio of America, Ltd.)*

and maintenance controls are provided on stereo amplifiers. These controls comprise:

1. Input selector switch
2. Bass level control
3. Treble level control   *to provide an equalizing characteristic for*
4. Tone control switch and low-boost switch   *the freq. response*
5. Low-cut filter
6. High-cut filter
7. Tape-dubbing switch
8. Volume control and touch-mute switch
9. Bass frequency selector switch
10. Treble frequency selector switch
11. Input impedance adjuster for phono–1
12. Input sensitivity control for phono–1 and 2
13. Mode selector switch
14. Tape monitor switch
15. Attenuator volume
16. Speaker selector switch
17. Balance control
18. Power switch

Although disc recordings are equalized in accordance with Record Industry Association of America (RIAA) standards, overall tonal balance may vary from one disc to another. The "new" RIAA equalization curve is shown in Figure 1–18. Differences in acoustic environments may also require subtle degrees of tonal compensation; because of the wide range and overlapping crossover characteristics of these requirements, conventional tone controls may be inadequate. In turn, an advanced stereo preamplifier provides a linear equalizer control for supplementary tonal compensation; it subtly augments conventional tone controls. Alternatively, a separate equalizer may be used, as pictured in Figure 1–19. An equalizer provides extensive control of amplifier frequency characteristics; it can compensate for room acoustics as well as differences in turntable cartridges and in speakers. Equalizers can also be used effectively with tape recorders to filter out noise from deteriorated records or from marginal broadcast signals. Equalizer frequency responses are illustrated in Figure 1–20. An advanced type of equalizer is termed the *parametric equalizer*. This design provides control of the center frequency for boost or cut, adjustment of bandwidth, and control of amplitude level. A parametric equalizer is usually designed with operational amplifiers (op amps) that provide precise control of frequency characteristics and also avoid the insertion loss that is imposed by passive types of equalizers.

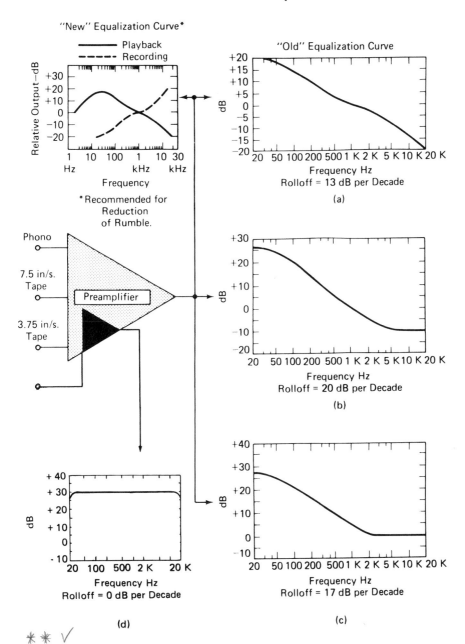

Figure 1–18 Examples of standardized preamp input frequency characteristics. (a) RIAA equalization curve for playback of records; (b) NAB standard
playback curve for 7.5 in/s. tape; (c) MRIA playback curve for 3.75 in/s.
tape; (d) tuner frequency characteristic is flat and amplifier operates at reduced gain.

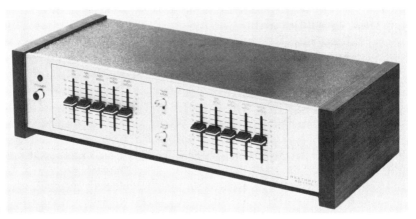

Figure 1–19 Appearance of a stereo equalizer. *(Courtesy, Heath Co.)*

Amplifiers used in advanced stereo systems are often equipped with audio-muting circuitry that permits instantaneous attenuation of the main volume level so that the listener can temporarily mute the system if he wishes, and then resume the same main volume level of operation. A slight touch of the touch-mute control reduces the main volume level by 20 dB, or to one tenth of its original intensity. Normal level can be resumed by touching the outer metal portion of the control. A volume control often has a logarithmic attenuation characteristic, which is an advantageous taper since the ear has a logarithmic loudness characteristic. A logarithmic taper in a volume control provides an adjustable loudness level proportional to its degrees of rotation.

Consider the operation of an input impedance adjuster in a preamplifier. In Figure 1–21, this control is connected to the phono-1 input terminals. Except for one special type of low-impedance phono cartridge, nearly all cartridges of the moving-magnet (MM) type are rated for a load impedance of approximately 50 kΩ. Variations in load impedance values for the cartridge affect its frequency response considerably. If the load impedance is low, a treble cut results. On the other hand, if the load impedance is high, a peak response is introduced into the treble range. Because varying load values affect phono cartridges with high internal impedances more than in cartridges with low internal impedances, an input impedance adjuster for a preamp can optimize system operation for various kinds of cartridges. In this example, the adjuster provides an input-impedance range of 30 kΩ to 100 kΩ.

In addition, an input sensitivity adjuster provides flexibility in operation with various kinds of phono cartridges. These magnetic, photo-electric, electrostatic, and piezoelectric designs are referred to as moving-

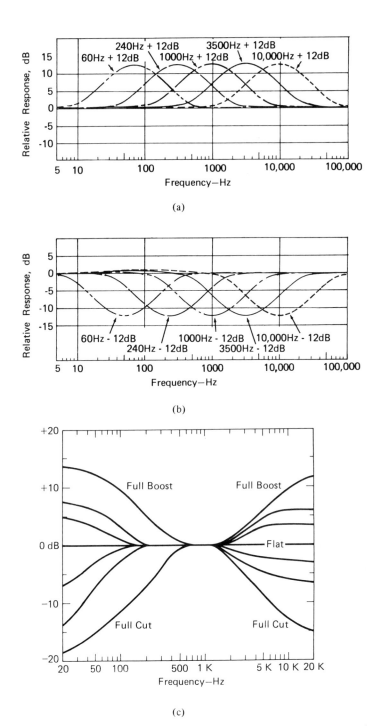

Figure 1–20 Typical equalizer frequency responses. (a) Maximum boost characteristics; (b) maximum attenuation characteristics; (c) comparative tone-control frequency responses.

23

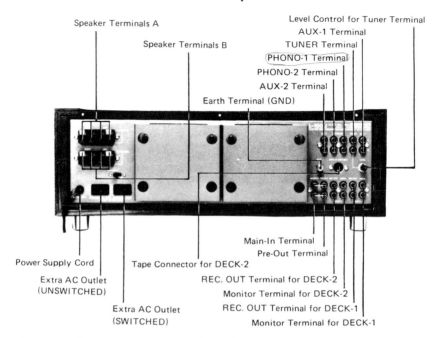

Speaker Terminals A

Speaker Terminals B

Level Control for Tuner Terminal
AUX-1 Terminal
TUNER Terminal
PHONO-1 Terminal
PHONO-2 Terminal
AUX-2 Terminal
Earth Terminal (GND)

Main-In Terminal
Pre-Out Terminal

Power Supply Cord   Tape Connector for DECK-2

Extra AC Outlet   REC. OUT Terminal for DECK-2
(UNSWITCHED)   Monitor Terminal for DECK-2

Extra AC Outlet   REC. OUT Terminal for DECK-1
(SWITCHED)   Monitor Terminal for DECK-1

Figure 1–21 Rear view of amplifier depicted in Figure 1–17. *(Courtesy, Lux Audio of America, Inc.)*

magnet (MM), moving-iron (MI), induced magnet (IM), and moving-coil (MC) types respectively. In Figure 1–21, the phono terminals of the amplifier match these various cartridges; however, a cartridge that has a low output level (from 0.01 to 0.1 mV) cannot be directly connected to the phono terminal. In turn, input sensitivity adjustment is required at both the phono–1 and phono–2 terminals. This example illustrates a continuous adjustment over a ± 5 dB range, with a level of 3 mV stipulated as zero dB. The most appropriate sensitivity value for a cartridge falls between 1.7 mV and 5 mV, as provided by the input sensitivity adjuster. Advanced stereo preamps often include scratch and rumble filters in the phono input channel (Figure 1–22). A scratch filter reduces the noise from deteriorated records at the expense of high-frequency response; a rumble filter reduces the audibility of mechanical vibration and other low-frequency interference at the expense of bass response.

Various advanced stereo preamps also provide a presence control (Figure 1–23) that can be switched in or out. A presence control introduces a regional rise in midrange frequency, peaked at approximately 2 kHz. Some listeners find that this "coloration" of the signal

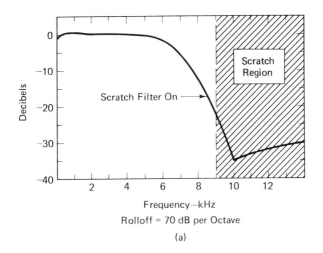

Frequency—kHz

Rolloff = 70 dB per Octave

(a)

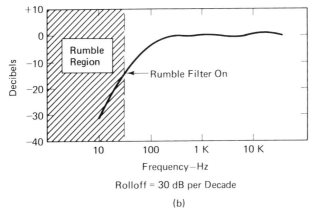

Frequency—Hz

Rolloff = 30 dB per Decade

(b)

Figure 1–22 Scratch and rumble filter characteristics. **(a)** Frequency response for a scratch filter; **(b)** frequency response for a rumble filter.

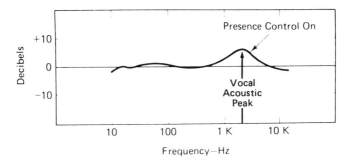

Figure 1–23 Amplifier frequency characteristic with presence control turned on.

enhances the natural sound of vocal selections and produces an illusion of intimacy.

Consider next the nonlinear relation of loudness (phon) units to relative power (decibel) units, as shown in Figure 1–24. The sensitivity of the human ear for bass tones is seen to drop off rapidly at low volume levels; ear sensitivity for treble tones also drops off to some extent at low volume levels. In order to obtain "natural" sound reproduction at less than normal acoustic levels, advanced stereo preamps generally include a loudness control or a loudness switch to supplement the volume control. The loudness control has a frequency characteristic that is opposite to that of the phon-dB curves depicted in Figure 1–24. As a result, the listener can adjust the loudness control for optimum "naturalness" at any setting of the volume control.

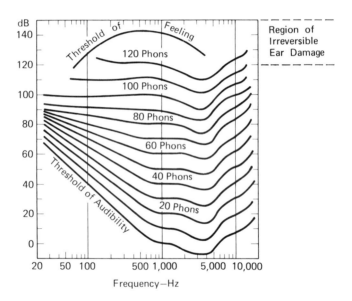

Figure 1–24 Relation of phon (loudness) units to decibel (relative power) units.

Stereo power amplifiers generally have a balance control that can increase and decrease the relative sound-level intensities from the L and R channels. A balance control may be set to its midpoint, but stereo reproduction often improves at an off-center setting. Most balance controls provide a very gradual change in L and R level relations over the first 90 degrees on either side of midpoint position. This charactertistic facilitates small changes in relative level adjustment. However, level variation is rapid over the remainder of the balance-control range; one channel will be completely silenced and the other channel will have maximum output at either extreme of control rotation. Different speaker efficiencies or acoustic asymmetry in the listening area may necessitate balance adjustments. Moreover, recordings and program material do not always have an ideal balance between their L and R channel signals.

VOLUME EXPANSION

A few advanced stereo amplifiers include a built-in volume-expansion section. In most designs, however, an auxiliary expander has to be connected between the preamplifier and the power amplifier (Figure 1–25) if the audiophile desires a volume-expansion function. Volume expansion denotes nonlinear amplification in which the output voltage increases more rapidly than the input voltage, as depicted in Figure 1–26. Conversely, volume compression denotes nonlinear amplification in which the output voltage increases more slowly than the input voltage, as exemplified in Figure 1–27. Volume compression may be used in recording procedures to reduce the dynamic range of the audio

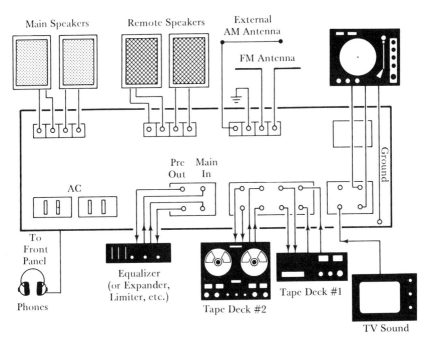

Figure 1–25 A volume expander may be connected at the preamplifier output terminals.

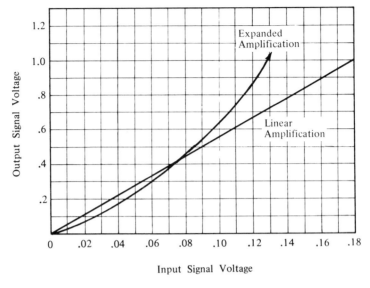

Figure 1-26   A comparison of linear amplification with expanded amplification.

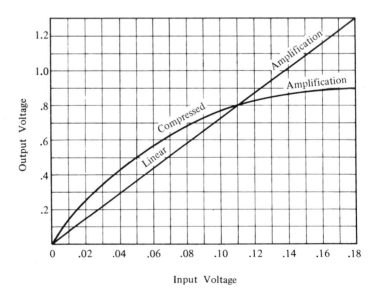

Figure 1-27   Comparison of linear amplification with compressed amplification.

signal and thereby avoid distortion due to saturation of the tape; volume compression may also be used to avoid overcutting of grooves in disc recordings. Broadcast program material is often compressed to avoid distortion due to overmodulation. Although compression does not deteriorate sound reproduction seriously, high-fidelity reproduction requires that compressed material be correspondingly expanded.

An example of an expanded waveform is shown in Figure 1–28. Observe that the amplitude intervals become progressively greater as the signal level increases. Consider a musical passage that has been substantially compressed; its dynamic range is correspondingly reduced and the passage will not sound entirely realistic to a musician. However, if the recorded passage is suitably expanded and then reproduced, its dynamic range will have been restored and the reproduced sound will seem realistic to the musician. After a musical passage has been compressed, its dynamic range is reduced at the expense of more or less harmonic distortion. In other words, nonlinear amplification is inevitably associated with harmonic distortion. However, if a compressed musical passage is then proportionally expanded, the harmonic

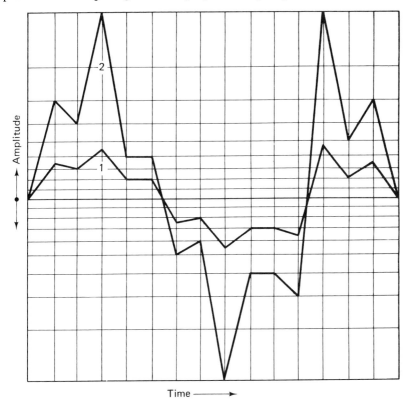

Figure 1–28   Example of an expanded waveform.

distortion associated with the expansion will cancel out the distortion associated with the compression. Not only has the dynamic range of the musical passage been corrected, but the incidental harmonic distortion has been cancelled.

Note in passing that any amplifier has a certain dynamic range which, expressed in dB, denotes the amplifier's signal-handling capability. In Figure 1–29, maximum power output from the amplifier is specified as the level at which harmonic distortion rises to 10 percent. Conversely, minimum output power from the amplifier is specified as the level at which the signal-to-noise ratio is 20 dB. In other words, the noise power is 20 dB less than the signal power at the noise "floor." Thus, the dynamic range of an amplifier is equal to the ratio, stated in dB units, of audio power at the distortion "ceiling" to audio power at the noise "floor."

Dynamic range =

$$10 \log \frac{Power \ at \ ceiling}{Power \ at \ 20dB \ SNR}$$

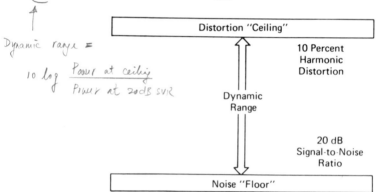

Figure **1–29**   Significance of the rated dynamic range for an amplifier.

A dynamic range of 40 dB is generally considered to be adequate for a hi-fi system. An advanced stereo system may achieve a dynamic range of 70 dB. Observe carefully that this 70–dB value refers to a distortion ceiling of 10 percent, whereas the high-fidelity distortion ceiling necessarily has a reference value of 1 percent. In turn, the high-fidelity dynamic range of an amplifier is somewhat less than its rated dynamic range. This difference in dynamic ranges is usually less than might be supposed, inasmuch as the harmonic distortion of an amplifier increases very rapidly as its region of overdrive is approached.

TURNTABLE CONSIDERATIONS

Turntables are grouped into automatic and single-play (manual) types. Either design is capable of advanced performance. Automatic turntables designed as changers have a spindle that typically holds a stack of six records; the records are played sequentially without attention from the operator. A typical changer appears in Figure 1–30, and basic nomenclature in Figure 1–31. Various kinds of drive mechanisms and control arrangements can rotate the turntable as long as they maintain constant speed. Variation in speed results in distortion, termed wow and/or flutter. Wow denotes a comparatively slow speed variation, whereas flutter denotes a rapid variation. Either change in speed produces frequency modulation of the recorded tones; both pitch and tempo become varied and the resulting impairment of fidelity is quite unpleasant to the listener. Advanced stereo systems may employ a DC servomotor type of turntable. This design has a quartz crystal frequency standard with phase-locked loop (PLL) circuitry that constantly monitors the platter speed and corrects for any fluctuation.

Figure 1–30   A good-quality type of turntable. *(Courtesy, Radio Shack)*

Rumble filters, described above, will not be needed in a turntable that qualifies for an advanced stereo system because its rumble value is below the threshold of audibility. Note that rumble is likely to be associated with wow; its source is in the mechanical system from drive motor to platter. As depicted in Figure 1–32, stroboscopic facilities can check or monitor turntable speed and can detect the presence of wow or flutter. Modern turntables often include a built-in strobe monitor that operates on the rim of the platter. When the platter speed vernier is adjusted correctly, the strobe lines or dots appear to be stationary. If the speed is substantially incorrect, the lines or dots will

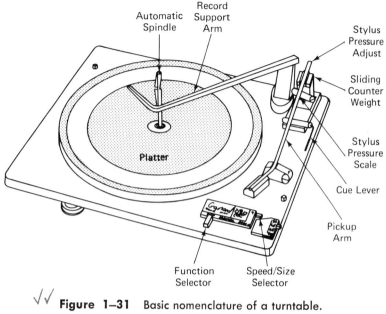

√√ **Figure 1–31**   Basic nomenclature of a turntable.

be perceived as a blur. In the case of wow, the lines or dots will appear to be stationary for a brief period, and will then appear to move clockwise. Soon this motion slows down; again the lines or dots briefly appear to be stationary, and then they start moving counterclockwise. This cycle repeats indefinitely. Flutter is associated with the same symptom, except that the cycle is completed rapidly.

Wow and flutter can be minimized to some extent by a heavy and well-balanced platter, which tends to oppose change in speed of rotation because of its considerable inertia (flywheel effect). Of course, there is a practical limit to the weight of platter in a turntable, so that appropriate design must also be considered in the attempt to minimize torque fluctuation. A well-balanced platter ensures that centrifugal stress will not be a source of wow. A turntable suitable for an advanced stereo system has a pickup arm (tone arm) that produces very low friction and very small tracking error. A high-quality turntable will also include an antiskating arrangement to minimize distortion from tracking error and to reduce the required stylus pressure. Note that a few turntables have a radial pickup arm instead of a tangential pickup arm; the former, inherently free from skating forces, does not require an antiskating arrangement. The chief disadvantage of a radial pickup arm is its comparatively high cost.

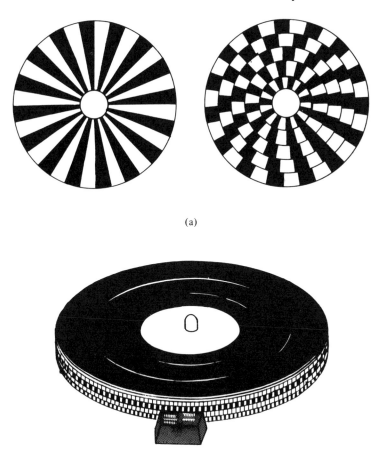

(a)

(b)

Figure 1–32   Stroboscopic facilities. (a) Strobe disks provide test of turn-table speed, and disclose wow or flutter; (b) built-in strobe monitor operates on rim of platter.

TAPE MACHINES

Tape machines are grouped primarily into open-reel and cartridge types. Although both are capable of high-fidelity performance, the open-reel design is preferred by some advanced stereo audiophiles. In a strict technical sense, a cassette is a miniature reel-to-reel tape package, whereas a cartridge is a miniature single-reel tape package. A cassette tape deck is illustrated in Figure 1–33. A tape recorder provides both recording and playback facilities, whereas a tape player lacks recording

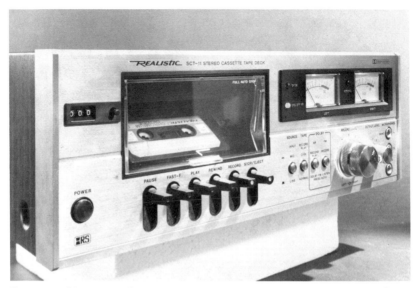

Figure 1–33   A good-quality cassette tape deck. *(Courtesy, Radio Shack)*

facilities. A tape deck lacks a built-in amplifier and is operated with an external amplifier and speaker system. Tape decks may or may not include recording facilities. Much of the background noise (principally hiss) that is audible in tape sound reproduction is present during the recording and manufacture of the tape. Chromium-dioxide tapes have a slightly lower noise level than ferric-trioxide tapes; they also have a slightly extended high-frequency capability. Advanced stereo systems generally include the Dolby* noise-reduction arrangement to minimize background noise. An active audio processor combines this noise-reduction circuitry with a volume-expander section for optimal reproduction of tape-recorded sound.

When a sound source has audible high-frequency noise, recording through a noise-reduction system as well as playing back through a noise-reduction system are advantageous, as depicted in Figure 1–34. Two Dolby sections in this arrangement permit instant playback while both the record and playback noise-reduction sections are operating. This mode of operation is termed A-B monitoring. It requires three heads in the tape machine. Most designs have control facilities to accommodate either ferric-trioxide tapes or chromium-dioxide tapes. Note that a machine designed solely for one type of tape will not provide acceptable reproduction with the other type of tape. Occasionally,

---

* Dolby is a registered trademark of Dolby Laboratories, Inc.

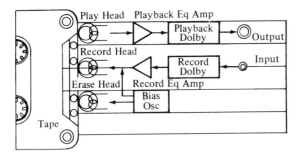

Figure 1–34   Skeleton arrangement of the Kenwood double-Dolby system.

a fourth head is included in a tape recorder for optimum bias application; this elaboration can improve high-frequency response somewhat and is of interest to advanced stereo enthusiasts. The fourth head also helps to reduce tape hiss.

2

Advanced Stereo Acoustics

BASIC PRINCIPLES

Symphony orchestras can perform in concert halls designed for nearly optimum acoustic characteristics, whereas the stereophile must endeavor to reproduce symphonic sound in a comparatively small listening area that inevitably "colors" the musical passages. In other words, the reverberation time of a small listening area is less than ideal. Room resonances are occasionally troublesome and the treble sound field in particular may be highly irregular. The bass sound field has comparatively high intensity in the eight corners of a rectangular listening area, as depicted in Figure 2–1. Although acoustic conditions in a particular listening area may modify the following basic rule, bass-tone intensity will probably be three times greater in any corner of a room than at the center. Conversely, treble tone intensity will be comparatively low in any corner of the room.

Consider the radiation of sound waves by a speaker in its axial direction, as shown in Figure 2–2. A sound wave is a longitudinal wave consisting of regions of alternate condensations (high pressure) and rarefactions (low pressure). These regions travel out from the sound source at a speed of 1130 ft/sec. As shown by the dots in the diagram, air molecules merely oscillate back and forth about their average position—only the compressions and rarefactions speed away from the sound source. The distance from one compression to the next along the line of propagation is equal to the length of the sound wave. In turn, the frequency of the wave is equal to the speed of sound divided by the wavelength. For example, a 1–kHz sound wave has a length of 1.1 feet, as indicated in Figure 2–3. The amplitude or intensity

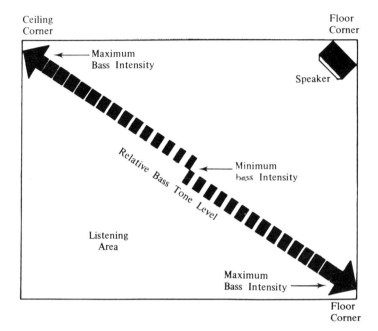

Ceiling
Corner

Floor
Corner

Maximum
Bass Intensity

Speaker

Relative Bass Tone Level

Minimum
bass Intensity

Listening
Area

Maximum
Bass Intensity

Floor
Corner

√ **Figure 2–1**   Corner-to-corner variation in bass-tone intensity.

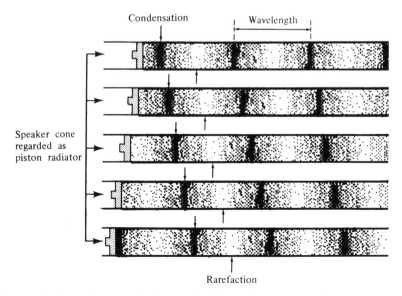

Condensation

Wavelength

Speaker cone
regarded as
piston radiator

Rarefaction

√ √
Figure 2–2   Radiation of a longitudinal sound wave at 45-degree intervals
by a piston speaker.

Advanced Stereo Acoustics

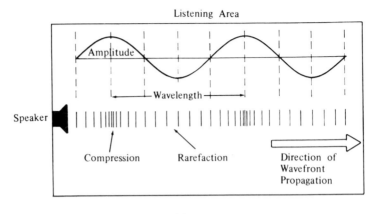

(a)

Frequency Hz	Wavelength Feet
20	55
100	11
1 K	1.1
10 K	0.11
20 K	0.05

(b)

Figure 2–3 A sound wave consists of successive regions of compression and rarefaction. **(a)** Representation of a sound wave in a listening area; **(b)** wave-length-frequency relations.

of a sound wave is equal to its degree of compression and rarefaction and is measured at any point within a listening area by means of a sound-level meter calibrated in dB units.

When a sound ray strikes a wall within a listening area, the ray is reflected as depicted in Figure 2–4: its angle of incidence is equal to its angle of reflection. Not all of the energy in the sound wave will be reflected, because some will be absorbed by the wall, depending upon its acoustic absorption coefficient. Accordingly, the energy level in the sound ray rapidly decreases as the ray is successively reflected from the walls within the listening area. Acoustic absorption coefficients for common substances are tabulated in Table 2–1. These coefficients are directly related to the reverberation time of a listening area. This is

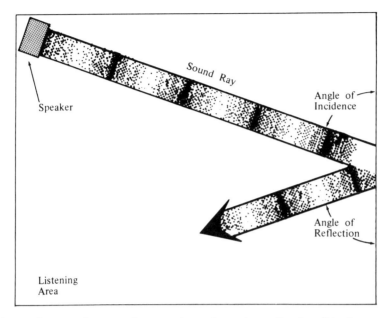

Figure 2–4   Reflection of a sound ray from the walls of a listening area.

the time that sound takes to decay to one-millionth of its original intensity, or to decay through a range of 60 dB. An "average" acoustic environment has a reverberation time on the order of 0.6 second. This reverberation time can be increased or decreased by suitable acoustic treatment of the listening area. The same musical passage will sound "dry" if the reverberation time is very short; conversely, if the reverberation time is excessively long, the musical passage will be reproduced with a "cathedral effect." A "dry" acoustic environment is seldom preferred, except for some forms of electronic music. On the other hand, a "cathedral effect" is ordinarily desired only for associated classes of pipe-organ compositions.

An advanced stereo system should include some method of acoustic environmental control. As indicated in a later section, sliding draperies and various types of acoustic screens may help to control reverberation time. Note that the reverberation time of a listening area is not directly related to the sound-intensity patterns within the area. In other words, one listening area in which treble tonal energy is distributed almost uniformly can have the same reverberation time as another listening area in which treble tonal energy is highly concentrated at the center of the room space. Again, this distribution depends upon the acoustic treatment of the listening area.

TABLE 2–1

ACOUSTIC ABSORPTION COEFFICIENTS
FOR VARIOUS FREQUENCIES

Substance	Absorption Coefficients					
	125 Hz	250 Hz	500 Hz	1,000 Hz	2,000 Hz	4,000 Hz
Brick, glazed surface	.03	.03	.03	.04	.05	.07
Carpet, heavy, on concrete floor	.02	.06	.14	.37	.60	.65
Carpet, heavy, on 40-ounce hairfelt or foam rubber padding	.08	.24	.57	.69	.71	.73
Concrete block, coarse surface	.36	.44	.31	.29	.39	.25
Concrete block, painted surface	.10	.05	.06	.07	.09	.08
Fabric, medium velour, 14 ounces per square yard, draped to half of its flat, undraped area	.07	.31	.49	.75	.70	.60
Floors						
concrete or terrazo surface	.01	.01	.015	.02	.02	.02
linoleum, asphalt, rubber, or cork tile on concrete	.02	.03	.03	.03	.03	.02
wood	.15	.11	.10	.07	.06	.07
Glass						
large panes of heavy plate glass	.18	.06	.04	.03	.02	.02
ordinary window glass	.35	.25	.18	.12	.07	.04
Gysum board, ½ inch, nailed to 2 inch x 4 inch lumber, centers 16 inches apart	.29	.10	.05	.04	.07	.09
Marble or glazed tile surface	.01	.01	.01	.01	.02	.02
Plaster, gypsum or lime, rough finish on lath	.02	.03	.04	.05	.04	.03
Plywood paneling, ⅜ inch thick	.28	.22	.17	.09	.10	.11
Water surface, as in a pool	.008	.008	.013	.015	.02	.025

√√ **Note:** A typical living room will have an acoustic absorption coefficient of approximately 0.25 at 500 Hz. In other words, one-quarter of the energy in a sound ray is absorbed, and three-quarters of the energy is reflected.

SPEAKER LOCATION

A modern concert hall is designed for optimum acoustic characteristics so that when stereo speakers are to be installed in a concert hall, they should be placed at the front and to the left and right of the orchestra position, as shown in Figure 2–5. On the other hand, similar placement of speakers relative to sound source in a residential listening area could leave much to be desired from an acoustical viewpoint because few residences are designed with any consideration for acoustics. As

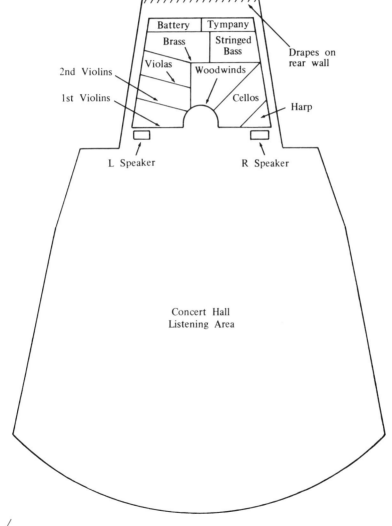

Figure 2–5   Correct placement of stereo speakers in a concert hall.

a result, the most satisfactory positions for stereo speakers in a residential listening area often have to be determined by experiment. Note in Figure 2–5 that no two wall surfaces are parallel in the concert hall. Eliminating parallel reflecting surfaces minimizes the intensity of sonic standing waves and helps to develop a uniform sound field at all audio frequencies. Most residential listening areas have parallel walls that tend to develop high-amplitude standing waves at certain audio frequencies, unless appropriate acoustic treatment is employed.

It is helpful to consider basic placements of a single speaker in a rectangular listening area with walls, floor, and ceiling that have the same acoustic absorption coefficient, and to observe the relationship between these locations and the reproduction of bass and treble tones. First, note the speaker placement depicted in Figure 2–6; thus, when a speaker is placed on a table in the middle of a rectangular room, bass tones develop minimum intensity throughout the listening area. Although this effect is modified slightly in nonuniform and less-than-ideal rectangular listening areas, it occurs in most residential listening areas. Therefore, a useful rule-of-thumb is to experimentally place a speaker in the middle of a rectangular room. Second, if the table is removed in Figure 2–6, and the speaker is placed on the floor, intensity of bass tones within the listening area will increase by approximately 40 percent. Third, if the speaker is placed against a side wall, as depicted in Figure 2–7, the bass tone intensity within the listening area will be almost double the intensity in the situation shown in Figure 2–6.

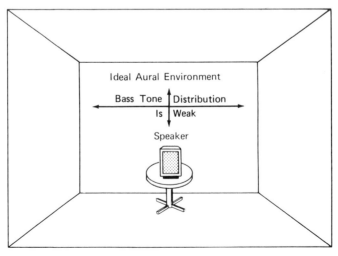

Ideal Aural Environment

Bass Tone Distribution

Is Weak

Speaker

✓ **Figure 2–6**   A speaker placed on a table in the middle of a room develops minimum bass output.

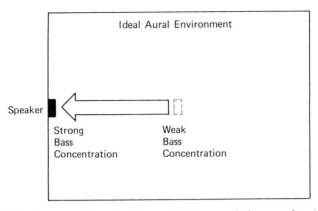

Figure 2–7 Bass-tone intensity is greatly increased if a speaker is moved from the center of a room to a side wall.

Finally, if the speaker is placed in a corner of the room, as pictured in Figure 2–8, the bass tone intensity within the listening area will be approximately triple the intensity of the situation depicted in Figure 2–6. This perhaps unexpected increase in bass tone output results from the semihorn action provided by the floor-wall-wall surfaces. Note also that the relative intensity of bass tones perceived by the listener will depend upon his listening position. For example, if the listener stands in the center of the room, bass tones will not sound as loud as when he moves to one of the walls. Bass tones will sound loudest when the listener moves into one corner of the room. To summarize briefly, the effective low-frequency characteristics of a speaker can be modified both by speaker placement and by listening position.

Any acoustic environment other than a completely "dead" or anechoic room will produce sound coloration. In Figure 2–9, the speaker is placed at floor level at the center of the front wall. As the listener moves his position from directly in front of the speaker toward the rear of the room, the sound quality (coloration) changes progressively. Note that minimum reverberant energy is perceived when the listener is directly in front of the speaker. On the other hand, a large proportion of reverberant energy is experienced when the listener is at the back of the room. Observe that if the listener uses earphones, he will hear only direct energy from the transducer, and the listening area will produce no sound coloration. When adequate acoustic treatment of a difficult residential acoustic environment is not feasible, earphones are mandatory in order to realize the capabilities of an advanced stereo system. Most acoustic environments can be satisfactorily controlled and optimized without undue cost and effort.

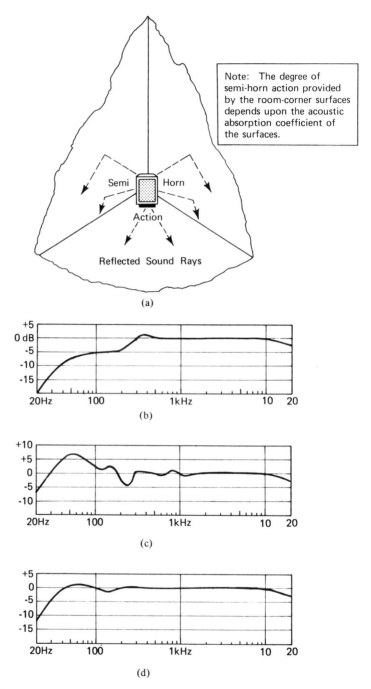

Note: The degree of semi-horn action provided by the room-corner surfaces depends upon the acoustic absorption coefficient of the surfaces.

(a)

(b)

(c)

(d)

Figure 2–8 Bass-tone intensity is maximized when a speaker is located in the corner of a room. (**a**) Semi-horn action; (**b**) example of speaker frequency characteristics; (**c**) room acoustic profile; (**d**) over-all frequency response of speaker/room system.

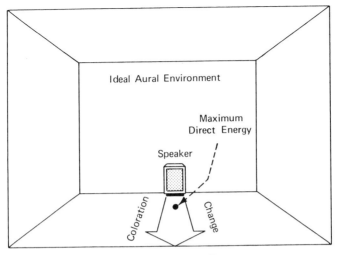

Ideal Aural Environment

Maximum
Direct Energy

Speaker

Coloration

Change

Maximum Reverberant Energy

Figure 2–9 Acoustic coloration varies as listener moves toward the back of the room.

SOUND MIRRORS AND PHANTOM AREAS

An acoustic image is analogous to an optical image, as shown in Figure 2–10. In this example, a tile wall reflects sound rays from a speaker to the listener. Tile has a very small acoustic absorption coefficient (very high reflectivity) and functions as a sound mirror. Accordingly, the listener experiences an illusion of another sound source behind the wall. He seems to be hearing sound from two sources: from the real speaker in front of him, and from the imaginary speaker (acoustic image) to his right. This acoustic image is an effective virtual sound source; the reflected sound rays serve to increase the sound level perceived by the listener. Reverberation in a listening area is the result of sound-wave reflections from the four walls, from the ceiling, and from the floor. In a typical acoustic environment, this reverberation increases the effective sound level by 6 dB. Observe that a real sound source in a listening area is associated with a virtual sound source in a phantom listening area, as pictured in Figure 2–11. This form of acoustic analysis is often helpful in evaluating the characteristics of a particular aural environment.

Consider a listening area such as the one depicted in Figure 2–12. One wall reflects sound waves whereas the opposite wall completely absorbs sound waves. In turn, the reflective wall is associated with a

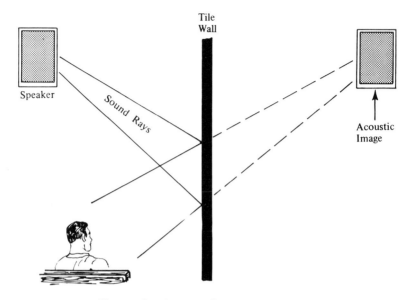

Figure 2–10   Sound mirror and image.

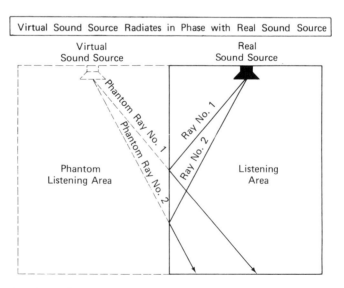

Figure 2–11   A phantom listening area contains a virtual sound source and is a "mirror image" of the real listening area.

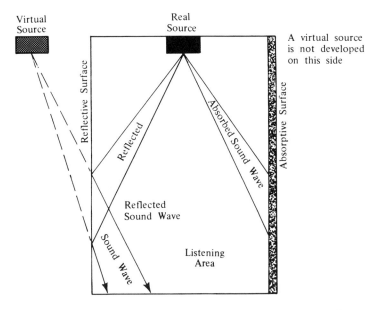

√ **Figure 2–12**   A completely absorbed sound wave has no virtual source.

virtual source of sound, whereas the absorptive wall does not develop a virtual source. The intensity of the virtual sound source is a function of the acoustic absorption coefficient for the wall between the real source and the virtual source. A conventional rectangular listening area has six reflecting surfaces. The four walls are associated with four principal phantom areas and four principal virtual speakers, as shown in Figure 2–13. Note that the fourth phantom area exists even if the speaker has a closed back. Although the speaker has no direct rearward radiation, the sound waves from the front of the speaker disperse around the speaker and are reflected from the wall behind. These reflections correspond to a fourth phantom area behind the speaker. The effective sound intensities from these four phantom areas are a function of the acoustic absorption coefficients for their associated wall surfaces.

In the example of Figure 2–13, the listener is positioned on the speaker axis and toward the farther wall. Note that the four wall phantoms remain fixed regardless of listening position. On the other hand, the perceived sound color changes as the listener moves about the room. Coloration changes result from variation in the proportion of direct and virtual sound that the listener intercepts as he moves about the room. The virtual sound is essentially a close-in echo; because it lags

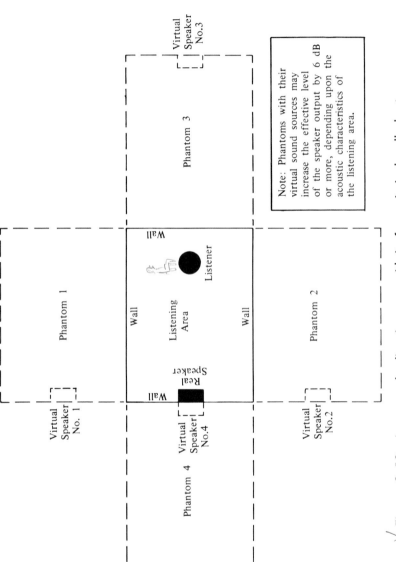

Note: Phantoms with their virtual sound sources may increase the effective level of the speaker output by 6 dB or more, depending upon the acoustic characteristics of the listening area.

Figure 2-13 A rectangular listening area with its four principal wall phantoms.

behind the direct sound in phase, the virtual sound changes the perceived audio waveform to some degree. Moreover, virtual sound is not a true "mirror image" of direct sound. Note in Table 2–1 that the acoustic absorption coefficients for various building materials are frequency selective. That means a typical wall surface absorbs more treble energy than bass energy. This is just another way of saying that the corresponding virtual sound source will be more or less "boomy" compared with the direct sound source. This frequency-selective action contributes significantly to the change in sound coloration perceived by the listener as he moves about the room.

In addition to the phantom areas pictured in Figure 2–13, a rectangular listening area is also associated with a floor phantom and a ceiling phantom, as depicted in Figure 2–14. These two phantom areas may contribute less to sound intensity and coloration than the four wall phantoms, because floors are often thickly carpeted and may have a high acoustic absorption coefficient. A sound-absorbent floor,

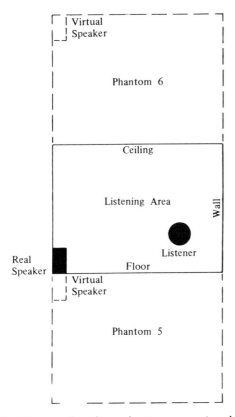

Figure 2–14 Floor and ceiling phantoms associated with a rectangular listening area.

which corresponds to a negligible floor-phantom area, leads to minimal reverberation between the floor and the ceiling, with the result that the ceiling phantom area contributes less to sound intensity and coloration than it would if the floor were uncarpeted. Of course, some rectangular listening areas have hardwood floors with few or no scatter rugs; this type of listening area is associated with a significant floor phantom. If the ceiling is also highly reflective, its corresponding phantom will have a substantial effect on perceived sound intensity and coloration.

Figure 2–15 shows that in addition to reflection of the direct sound rays from the speaker, there is ensuing reflection of the reflected rays. That is, first-order reflections are followed by weaker, second-order reflections. In turn, these second-order reflections correspond to second-order phantom areas and to second-order virtual speakers. Observe that the second-order phantom area (7) is a mirror image of the first-order phantom area (3), and that it is also a mirror image of the first-order phantom area (6). Although only one pair of sound rays is indicated in Figure 2–15, actually a dense population of rays is reflected from the ceiling and walls. It can be shown that this population has the same virtual-speaker sources that are indicated for a single pair of rays. Inasmuch as the reflected sound rays are re-reflected numerous times before they fall below the threshold of audibility, it follows that a listening area is associated with third-order phantoms, and so on. These higher-order phantom areas contribute comparatively little to the acoustic environment and can be ignored in a preliminary analysis.

OPEN-WINDOW UNITS

A directly-radiated sound wavefront from a speaker is propagated and reflected from the adjacent wall of the listening area, as shown in Figure 2–16 where the wall behind the speaker has uniform reflectivity over its entire surface, and both of the side walls have zero reflectivity. Accordingly, all acoustic reflection occurs from the wall behind the speaker, and the radiated sound-field intensity decreases as the square of the distance from the speaker. When the sound wavefront strikes the rear wall, its direction of propagation is reversed, as indicated in the diagram (the rear wall is reflective). If it is assumed that the rear wall has zero absorption, the sound wavefront reverses its direction of propagation without attenuation. Then, as the reflected wavefront travels back toward the speaker, its intensity decreases as the square of its distance from the rear wall.

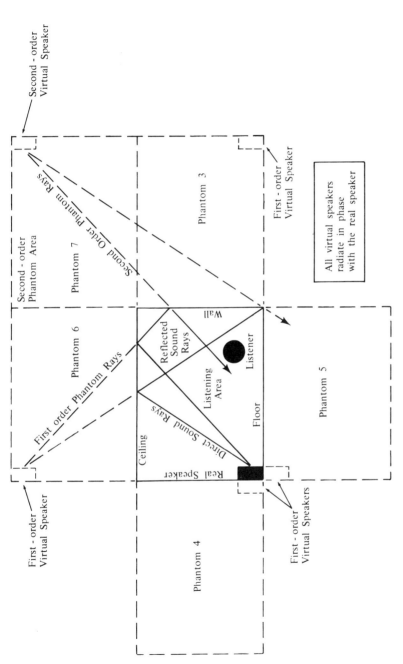

Figure 2–15 Second-order sound rays correspond to a second phantom area and relate to a separate virtual speaker.

Consider next the modified reflection process that occurs when there is an open window in the rear wall, as depicted in Figure 2–16(b). Not all of the direct wavefront is reflected because part of it strikes the open window and continues out into free space. The wavefront is reflected with a "hole" corresponding to the open window. In accordance with Huygens' principle,* every point in this reflected wavefront is to be considered the source of small secondary wavelets, which spread out in all directions with the speed of sound (1130 ft/sec). Consequently, progressive expansion of the reflected wavefront quickly "fills in" the initial acoustic "hole." Therefore, the reflected wavefront has reduced intensity in proportion to the area of the open window. Note that the acoustic absorption coefficients listed in Table 2–1 are termed open window (OW) units.

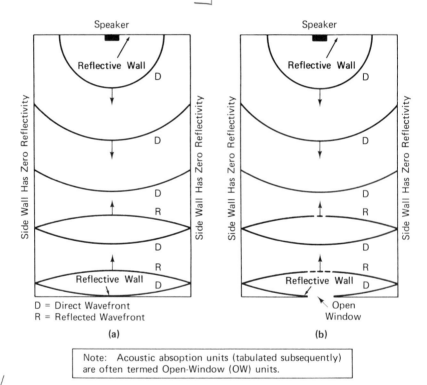

D = Direct Wavefront
R = Reflected Wavefront

(a)   (b)

Note: Acoustic absoption units (tabulated subsequently) are often termed Open-Window (OW) units.

Figure 2–16   Direct and reflected sound wavefronts. **(a)** Direct wavefront strikes rear wall and is reflected; **(b)** open window in rear wall attenuates reflected wavefront.

* Huygens, Christian: "Every point of an advancing wavefront is a new center of disturbance from which emanate independent wavelets whose envelope constitutes a new wavefront at each successive stage of the process."

STANDING WAVES AND ACOUSTIC RESONANCES

Sound reproduction within a listening area is colored by acoustic resonances that arise from standing waves, as exemplified in Figure 2–17. Resonances occur within any enclosure that has reflective surfaces, but are pronounced between parallel reflective surfaces. The resonant frequency of an enclosed space is determined by its dimensions. A standing wave has two components that propagate in opposite directions. In other words, one of the wavefronts retraces the path of the other wavefront; both wavefronts propagate at the speed of sound. At the resonant frequency, the incident wave and the reflected wave reinforce each other, and form a standing wave. There are many acoustic resonant frequencies for the opposite walls of a room, because any integral multiple of a wavelength that equals the distance between the parallel walls corresponds to an acoustic resonant frequency.

Note that the incident wavefront and the reflected wavefront move in opposite directions. Accordingly, a region of compression in the incident wavefront will be coincident with a region of compression in the reflected wavefront at half-wave intervals in the space between parallel walls. This is just another way of saying that the first resonant wavelength is equal to twice the distance between the parallel walls;

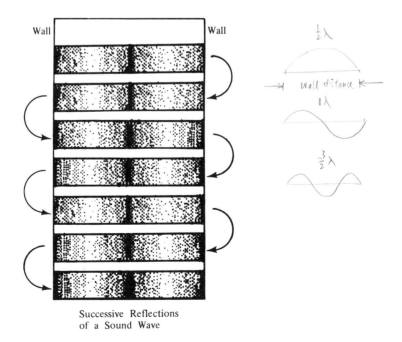

Successive Reflections
of a Sound Wave

Figure 2–17 Example of standing waves and acoustic resonance.

the second resonant wavelength is equal to the distance between the walls; the third resonant wavelength is equal to one-half the distance between the walls, and so on. Correspondingly, the first resonant frequency is equal to the reciprocal of twice the distance between walls; the second resonant frequency is equal the reciprocal of the distance between walls; the third resonant frequency is equal to the reciprocal of one-half the distance between walls, and so on. Note that the example of acoustic resonance shown in Figure 2–17 represents the second resonant wavelength, or second resonant frequency of the space between the two walls.

For a listener situated between two sound sources (such as two virtual sources), the intensity of perceived sound depends upon his position. At points of phase reinforcement, the sound field has maximum intensity; at points of phase cancellation, the sound field has minimum intensity. Phase reinforcement develops resonant points, whereas phase cancellation develops anti-resonant points (see Figure 2–18). However, the sound field does not drop to zero at anti-resonant points, and it does not attain its theoretical maximum at resonant points. Since the listening area is associated with phantom areas and virtual sound sources as depicted in Figure 2–19, resonant rise and fall in sound-field intensity tend to be smoothed out by the effects of the phantom rays. Inasmuch as the direct radiation has a higher intensity than the phantom radiation, resonant rise and fall in sound-field intensity cannot be completely smoothed out by this process.

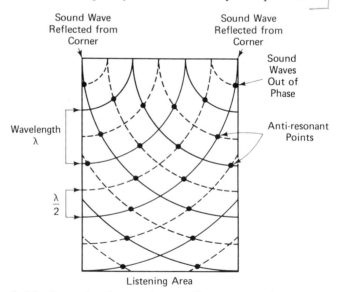

Figure 2–18 Example of acoustic cancellation points (anti-resonances) in a listening area.

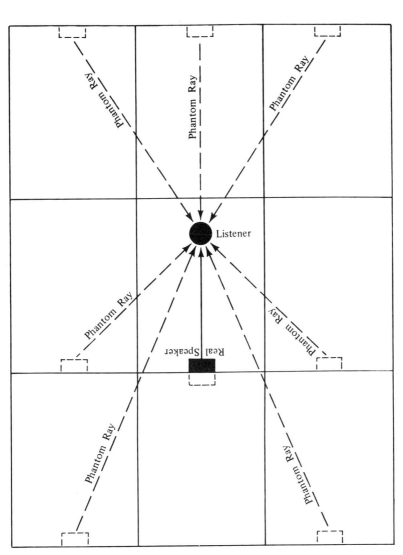

Figure 2–19 Resonant and anti-resonant variations in the sound field tend to be smoothed out by the effects of phantom rays.

REVERBERATION

Reverberation is an aspect of acoustic resonance; both involve sound energy storage. This storage increases the density of sound energy, with the result that a reproduced tone is perceived at a louder level than it would be otherwise. Reverberation is also associated with impairment of high-frequency reproduction, as shown in Figure 2–20. In this example, the time delay between the incident sound and the reflected sound causes the fifth-harmonic components in the two sound waves to cancel out. Accordingly, the listener perceives only the fundamental frequency. Shorter time delays result in cancellation of higher frequencies, and vice versa. If a sound wavefront undergoes numerous reflections with various time delays, only the low-frequency components of the original sound wave will be perceived by the listener. That is, reverberation in general is associated with a treble cut.

It follows that reverberation retards the growth and decay of the density of sound energy in the listening area, as exemplified in Figure 2–21. In these examples, two tone bursts with unequal durations are the source of acoustic wave envelopes in environments of zero, short,

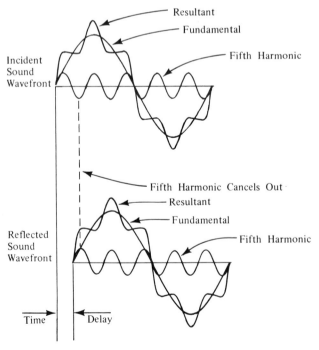

Figure 2–20 Time delay between incident and reflected sound wavefronts impairs high-frequency reproduction.

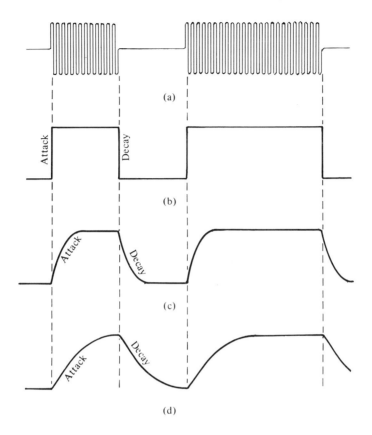

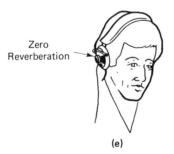

Figure 2–21 Effect of reverberation time on dispersion of acoustic energy. **(a)** Typical tone bursts, with unequal durations; **(b)** ideal growth and decay characteristic of acoustic envelope; **(c)** growth and decay of acoustic envelope with short reverberation time; **(d)** growth and decay of acoustic envelope with long reverberation time; **(e)** zero reverberation is provided by headphones.

and long reverberation times, respectively. With a zero reverberation time (as in an anechoic chamber), the attack (growth interval or rise time) is virtually instantaneous; similarly, the decay (fall time) is virtually instantaneous. Note that sound reproduction with zero reverberation time is ordinarily termed "unnatural" by casual listeners, who associate it with a feeling of annoyance and restlessness. Next, when the listening area has a short or moderate reverberation time, the attack and decay intervals of the acoustic envelope occupy a significant period of time (such as 0.75 second). This acoustic "distortion" is preferred by many listeners, who think that it imparts "mellowness" and "naturalness" to musical passages.

Although any finite reverberation time impairs articulation of speech passages to some extent, short reverberation time has little such effect. Observe next the growth and decay of an acoustic wave envelope in a listening area, as exemplified in Figure 2–21(d). If the reverberation time is on the order of 1.5 to 2 seconds, the listener describes the resulting sound reproduction as characteristic of "cathedral tones." In turn, this acoustic reproduction is appropriate only for musical compositions in a religious context. A long reverberation time will not greatly impair articulation, provided that speech passages are delivered slowly; on the other hand, fast delivery of speech passages will result in poor articulation. An optical analogy of reverberation pictured in Figure 2–22 shows reverberant sound analogous to multiple images. A verbal analogy is provided in Figure 2–23.

To recapitulate briefly, a controlled amount of reverberation contributes to the esthetic quality of vocal and musical passages. Among other factors, an optimum reverberation time is a function of the

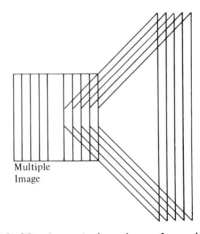

Multiple
Image

Figure 2–22   An optical analogy of reverberation.

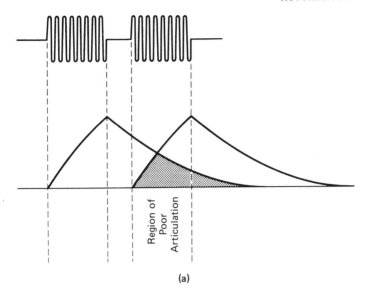

(a)

THE QUICK RED FOX JUMPED OVER THE LAZY BROWN DOG

versus

THEQEUICKRKEDFDOXJXUMPEDODVERTRHELEAZYBYROWNDNOG

(b)

Figure 2–23   Articulation of speech passages is impaired by an excessively long reverberation time. **(a)** Overlap of attack and decay intervals confuses the speech characteristics; **(b)** visualization of attack and decay overlap.

volume in a listening area. For instance, although a reverberation time of 0.75 second is generally preferred for a listening area that has a moderate volume, a large concert hall should have a longer reverberation time. Conversely, a small listening area, such as an automobile, should have a shorter reverberation time. Note in passing that although the reverberation time is zero for earphones, the reproduced sound is not altogether devoid of reverberation when earphones are used. That is, the reverberation characteristic of the recording studio or of the broadcasting studio will be present in the sound reproduced by earphones. A speaker installation in a listening area can be checked by first listening to a musical passage with earphones and then quickly switching to speaker reproduction. Speaker reproduction less esthetic than earphone reproduction indicates that the listening area needs acoustic treatment.

COLORATION OF REPRODUCED SOUND

Coloration of reproduced sound in a listening area occurs when the total path traveled by a sound ray from source to listener is less than 60 feet. Sound coloration is experienced because the listener is unable to distinguish between real and virtual sources in the close-in echoes. If, however, the total path exceeds 60 feet, an echo is perceived (the listener distinguishes between the real sources and virtual source). Coloration of reproduced sound is affected by reverberation time, resonances, anti-resonances, selective absorption, room volume, room shape, and reverberant attack and decay characteristics. An ideal acoustic environment would exhibit a basic exponential attack and decay characteristic; in practice, attack and decay deviate from exponential form (see Figure 2–24). Although a small listening area may have the same reverberation time as a large listening area (Figure 2–25), the listener will make a marked distinction between the two acoustic environments.

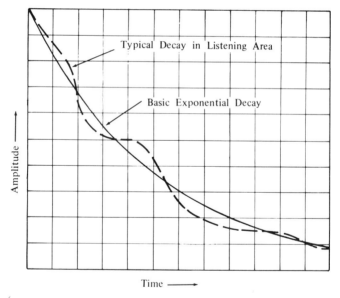

Figure 2–24 Basic versus typical decay characteristics.

Various resonant frequencies are associated with the different modes of reverberation in a listening area. If a hallway leads into the room, or if there are doorways from the listening area into adjoining rooms, sound coloration will also be a function of acoustic coupling. The relationship between physical coupling and beat development is

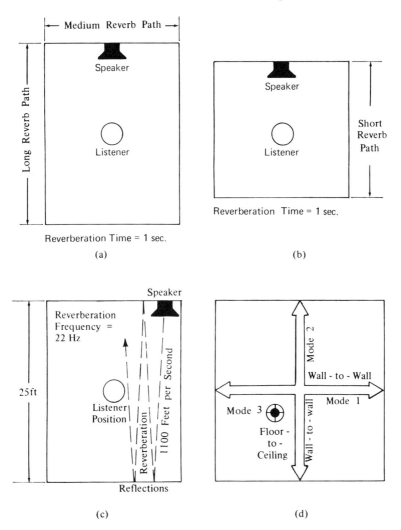

Figure 2–25 Basic effect of room volume on coloration of reproduced sound. **(a)** Large room has reverberation (echoes) at long intervals; **(b)** small room reverberates at short intervals; **(c)** front and rear walls are separated 25 feet, and this mode of reverberation has a frequency of 22 Hz; **(d)** examples of reverberation modes.

demonstrated by the pendulums in Figure 2–26. Since both pendulums are suspended from the same string, the period of the first pendulum affects the period of the second pendulum, and vice versa. The resulting beat frequency is equal to the difference between the periods of the two pendulums. Similarly, an open door between two rooms provides

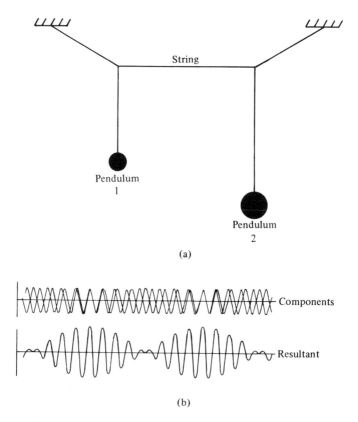

Figure 2–26   Examples of coupled pendulums and beat development. **(a)** Arrangement; **(b)** coupling characteristic.

acoustic coupling, so that the reverberation frequency of one room affects the reverberation frequency of the other. Values of coupling that produce audible beats develop a rise and fall in the sound-field intensity at a rate equal to the difference between the two reverberation frequencies. The effective amplitude of beats in acoustically coupled rooms can be reduced by provision of a direct-sound source in each room, as exemplified in Figure 2–27.

For all practical purposes, a residential acoustic environment is a linear system in which sound reproduction will not be colored by acoustic intermodulation or by acoustic harmonic distortion. However, an occasional exception may occur, as in the case of the wall construction feature depicted in Figure 2–28. Here, a large plastic sheet is loosely secured to a solid wall. Because the plastic sheet will vibrate when it is struck by sound waves, and because it makes localized and

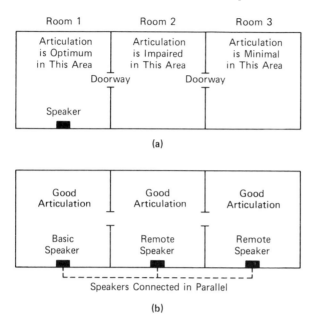

(a)

(b)

Figure 2–27 Articulation in acoustically coupled rooms. **(a)** Example of progressively poorer reproduction; **(b)** remote speakers provide good sound reproduction in acoustically coupled rooms.

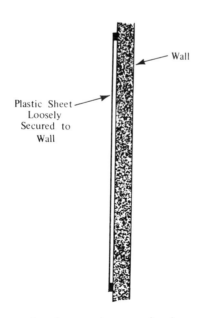

Figure 2–28 Construction feature that can develop acoustic intermodulation or harmonic distortion.

intermittent contacts with the wall, the resulting erratic mode of reflection is characterized by more or less harmonic or intermodulation distortion. This form of acoustic distortion can impart a highly disagreeable coloration to the reflected sound wave. The remedy is either to secure the plastic sheet firmly against the wall, or to provide sufficient spacing between the surfaces to prevent contact, even on the peaks of high-intensity sound wavefronts.

SOUND ENERGY DISTRIBUTION

Low-frequency sound energy diffracts extensively around most obstructions in a listening area. Thus, treble sound radiates only from the front of a closed-back speaker, whereas bass-tone intensity is almost the same both in front and back of the speaker, owing to the substantial diffraction that occurs. On the other hand, a marked treble cut occurs in the rear of the speaker because high-frequency sound energy tends to be "shadowed" by the speaker cabinet (enclosure). This screening effect becomes very prominent for common sound wavelengths that are shorter than the height or width of the obstruction. Accordingly, the furniture in a room is a dominant factor in establishment of its treble acoustic profile (see Figure 2–29). Reflection of treble wavefronts back and forth between parallel surfaces is also a major factor in this profile. As noted previously, most residential listening areas exhibit a considerably higher treble tone intensity in the central region of the room than in the corners of the room.

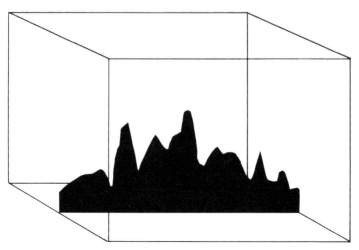

Figure 2–29   Treble acoustic profile of a listening area is generally quite irregular.

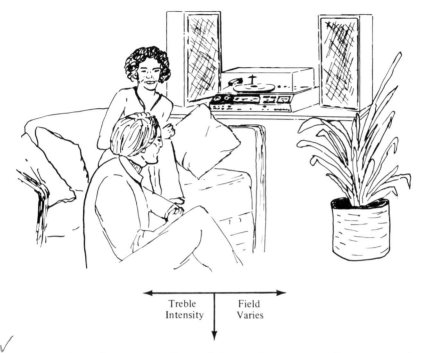

◄─── Treble │ Field ───►
    Intensity │ Varies
▼

Figure 2–30   The intensity of a 10-kHz sound wave may change as much as 10 dB when the listener moves 3 feet in a critical direction.

The treble profile of a typical residential listening area changes considerably when a person walks through the room, or when the listener moves a short distance in a critical direction (see Figure 2–30). For example, if the listener changes his position 3 feet in a certain direction, a sound-level meter may indicate that the intensity of a 10–kHz sound wave has changed as much as 10 dB. On the other hand, the listener may change his position 6 feet in another direction, and the intensity of the 10–kHz sound wave may remain virtually unchanged. Stated otherwise, high-frequency sound distribution in representative residential listening areas may be changed substantially by comparatively minor changes in acoustic reflection or screening at seemingly arbitrary positions. Therefore, the treble acoustic profile of a listening area must be experimentally determined with the aid of a sound-level meter such as the one illustrated in Figure 2–31.

Because treble tones are least diffracted and are reflected by comparatively small surfaces, the listener's perception of left (L) and right (R) sound directions in stereo reproduction is controlled chiefly by the higher audio frequencies (see Figure 2–32). Directionality clues

Figure 2–31   A sound-level meter is used in analysis of stereo sound fields. *(Courtesy, Radio Shack)*

are also provided by phase differences between sound rays from a source that arrive at the listener's left ear and at his right ear. In other words, the sound from an L speaker arrives at the listener's right ear slightly later than it arrives at his left ear. In turn, R sound perceived at his left ear lags behind the same sound perceived at his right ear. This phase difference contributes to the listener's directionality judgment. Also, an L sound ray that reaches his right ear is somewhat shadowed by the listener's head; consequently, the intensity of this ray is more or less attenuated and this intensity difference also contributes to the listener's directionality judgment.

Directionality judgments are affected by the absorbent and reflective characteristics of "problem" listening areas. As an illustration, Figure 2–33(a) depicts a room in which one wall is highly absorbent and the other surfaces are highly reflective. The listener perceives most of the sound from the L speaker as if it were arriving from the right of the listening position. Conversely, Figure 2–33(b) depicts a room in

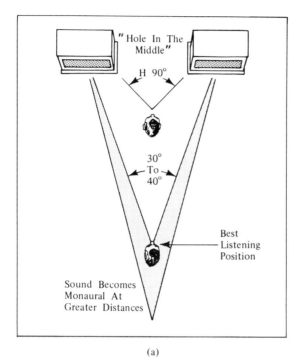

(a)

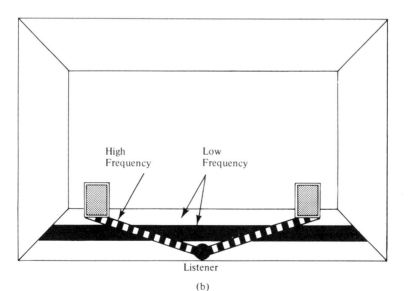

Listener

(b)

√√ **Figure 2–32** Directionality clues. **(a)** Directionality perception requires proper listening position; **(b)** directionality clues are provided chiefly by the high audio frequencies.

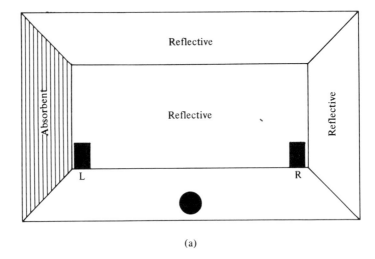

(a)

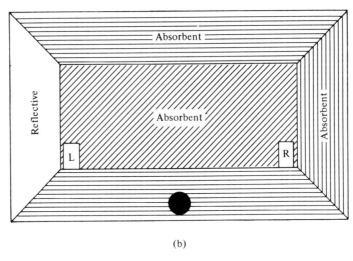

(b)

Figure 2–33 Two basic types of "problem" areas. **(a)** One surface highly absorbent, other surfaces highly reflective; **(b)** one surface highly reflective, other surface highly absorbent.

which one wall is highly reflective and the other surfaces are highly absorbent. In this acoustic environment, the listener perceives most of the sound from the R speaker as if it were arriving from the left of the listening position. Directionality clues can be improved by operating the R speaker at a somewhat lower level than the L speaker; however, because of the anomalous reflection characteristics of the acoustic environment, complete correction of directionality clues requires appropriate acoustic treatment of the listening area.

Another type of "problem" listening area is depicted in Figure 2–34, where open-cabinet construction at one end of the listening area introduces a distinct acoustic resonance. This resonant frequency produces a peak in the acoustic frequency response of the room and also has an abnormal reverberation time. At this resonant frequency, the listener experiences a "hangover effect" in the sound field. This cavity resonator can be regarded as a "monotone speaker" located in front of the open cabinet construction. This speaker reproduces the same information as the L + R signal, except that theoretically it is driven through a narrow bandpass filter. The bandpass filter is considered to have the same center frequency as that of the cavity resonance. Although this filter does not completely suppress off-resonance frequencies, it attenuates them substantially. This type of acoustic distortion can be corrected only by suitable acoustic treatment of the listening area.

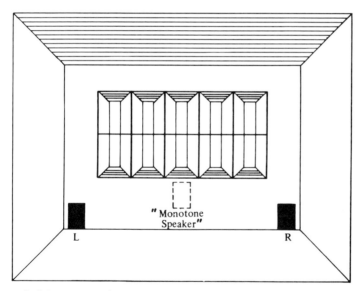

Figure 2–34 Open-cabinet structure produces an acoustic-resonant "problem" type listening area.

Still another type of "problem" listening area is pictured in Figure 2–35. This acoustic environment has a "sound trap" defect because the secondary listening area tends to absorb more sound energy from the L speaker than from the R speaker. Partial correction can be obtained by operating the R speaker at a somewhat reduced sound

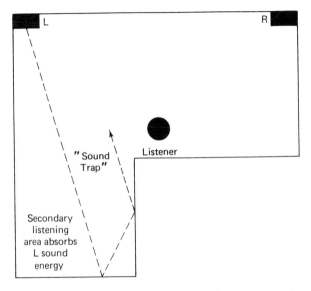

Figure 2–35 A secondary listening area absorbs more sound energy from one speaker than from the other.

level. Owing to asymmetrical reflective characteristics, however, the normal directionality clues are impaired. The secondary listening area also induces an abnormal acoustic resonance and tends to produce a peak in the acoustic frequency response of the room. In the event that the surfaces in the secondary listening area are highly reflective, a "hangover effect" may also intrude itself into the sound field. Although frequency equalizers for the L and R speakers may partly correct this problem, relevant acoustic treatment is needed for full correction.

3

Advanced Stereo Amplifiers

GENERAL REQUIREMENTS

Most high-fidelity amplifiers are very linear; that is, in other words, the amplifier output voltage is directly proportional to its input voltage. However, a recording amplifier may operate nonlinearly as a compressor and thereby reduce the dynamic range of an audio signal. Signal compression is used to avoid overcutting of grooves on phono discs, to avoid saturation of magnetic tapes, and to avoid overmodulation of radio transmitters. The compressed signal, however, should be correspondingly expanded before it is reproduced. If an audio signal has been substantially compressed, it will sound "squashed" when played back through a linear high-fidelity amplifier. Therefore, advanced stereo amplifiers are commonly equipped with a dynamic processor section that can expand a signal according to the compression that was used during the recording process. Thus, the dynamic processor section linearizes the recording/reproducing system. Note that a compression amplifier necessarily introduces some degree of harmonic distortion; similarly, an expansion amplifier introduces some degree of harmonic distortion. However, these are opposing distortion products, and in a properly designed compression/expansion system they cancel out and provide distortionless amplification, as illustrated in Figure 3–1.

Advanced stereo systems use class A, AB, B, D, G, and H amplifiers, as summarized in Table 3–1. Low-level stages are operated in class A; high-level stages are generally operated in class AB, D, G, or H. Modified class A stages commonly operate as both compressors and expanders; thus, a typical audio processor compresses small signal voltages that are near the noise level, and expands large signal voltages

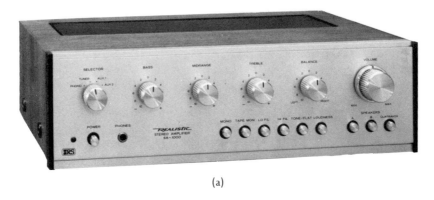

(a)

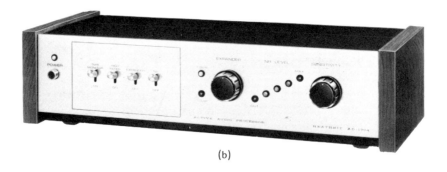

(b)

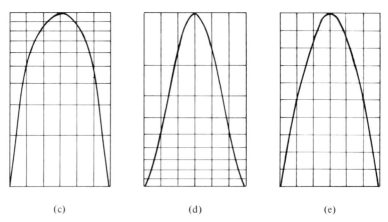

(c)   (d)   (e)

Figure 3–1   Stereo amplifiers and operating modes. **(a)** Appearance of a
high-performance stereo amplifier *(Courtesy, Radio Shack)*; **(b)** view of a
high-performance expansion amplifier *(Courtesy, Heath Co.)*; **(c)** section of
sine wave compressed; **(d)** section of sine wave expanded; **(e)** precise sine
wave obtained by compression (recording) followed by expansion (reproduction).

that are well above the noise level. This characteristic helps to reduce noise and restore normal dynamic range. Thus, this type of amplifier may compress very weak signals by 4 dB and expand very strong signals by 10 dB. This is just another way of saying that the amplifier attenuates a very weak (and noisy) input signal by 37 percent, and amplifies a very strong input signal three times. At approximately 20 mV input, such an amplifier neither attenuates nor steps up the input signal—input and output amplitudes are equal. Above 20–mV input level and up to 100 mV level, the amplification factor increases progressively from unity to three times. This variation in amplification factor is obtained by means of gain-control action, as depicted in Figure 3–2.

TABLE 3–1

AMPLIFIER CLASSES AND TRANSFER MODES

Class A	Linear Transfer Mode: Output waveform is the same as the input waveform (except for residual harmonic distortion) over 360 degrees of the operating cycle.
Modified Class A	Nonlinear Transfer Mode: An expansion amplifier develops an output waveform in which higher signal amplitudes are processed at higher amplification factors than are lower signal amplitudes.
Modified Class A	Nonlinear Transfer Mode: A compression amplifier develops an output waveform in which higher signal amplitudes are processed at lower amplification factors than are lesser signal amplitudes.
Class AB	Output waveform is the same as the input waveform (except for residual harmonic distortion) over 180 degrees of the operating cycle. A small amount of forward bias is used to minimize crossover distortion; class AB stages are operated in push-pull. Class AB amplifiers are usually designed with a linear transfer mode, but may be operated with expansion or compression transfer characteristics.
Class B	Output waveform is the same as the input waveform (except for residual distortion) over 180 degrees of the operating cycle. Class B stages are operated in push-pull. Class B amplifiers are usually designed with a linear transfer mode, but may be operated with expansion or compression transfer characteristics.
Class C	Output waveform is not the same as the input waveform. The output waveform occurs over less than 180 degrees of the operating cycle. Class C operation in a high-fidelity amplifier is a symptom of stage malfunction.

TABLE 3–1 *Continued*

Class D	Output waveform is the same as the input waveform (except for residual distortion) over 360 degrees of the operating cycle. Technically, a class D amplifier differs from a class A amplifier in that the input waveform is converted into a pulse-modulated form prior to amplification, and is then reconstituted into original form following amplification. Class D amplifiers are usually designed with a linear transfer mode, but may be operated with expansion or compression transfer characteristics.
Class G	Output waveform is the same as the input waveform (except for residual distortion) over 360 degrees of the operating cycle. A class G amplifier differs from a class A amplifier in that it has a bilevel type of operation. Signal amplitudes up to a medium level are amplified by the first-level section of the amplifier; signal amplitudes above the medium level are amplified by the second-level section of the amplifier. Outputs from the two sections are combined to reconstitute the original waveform. Class G amplifiers are usually designed with a linear transfer mode, but may be operated with expansion or compression transfer characteristics.
Class H	Output waveform is the same as the input waveform (except for residual distortion) over 360 degrees of the operating cycle. A class H amplifier differs from a class A amplifier in that it operates with a variable-level power supply. When processing a low-level signal, the amplifier operates with a low supply voltage; when processing a high-level signal, the supply voltage for the amplifier automatically increases as required in order to exceed the amplitude of the signal that is being processed. Class H amplifiers are usually designed with a linear transfer mode, but may be operated with expansion or compression transfer characteristics.
Magnetic Field Amplifier	Output waveform is the same as the input waveform (except for a small percentage of residual distortion) over 360 degrees of the operating cycle. It is a unique design that employs pulsed silicon controlled rectifiers instead of transistors. A typical magnetic field amplifier comprises two SCR's, four diodes, an AM modulator with control logic, and an optical isolator.

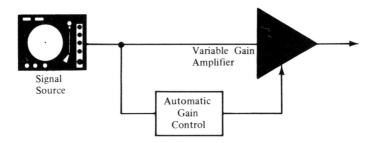

Figure 3–2 Functional block diagram for an advanced stereo nonlinear amplifier.

Preamplifiers in advanced stereo systems are evaluated primarily according to percentage harmonic distortion, phono hum level, and auxiliary input hum level, but numerous minor characteristics and features may also be taken into account. Power amplifiers are evaluated primarily according to percentage harmonic distortion, percentage inter-modulation distortion, power bandwidth, signal-to-noise ratio, and frequency response, along with optional considerations such as watts output per channel. Although music-power ratings are less meaningful than RMS power ratings, they provide an approximate measure of peak power output capability. Damping-factor values are less informative than tone-burst or square-wave response data, although they are related in a general way to system transient response.

AMPLIFIER ARRANGEMENTS

Basic classes of amplifier operation are depicted in Chart 3–1. Class–A stages are used chiefly in small-signal operation ranging from 1 μV to 10 mV. Class–A stages are also used in low-power operation up to 1 watt. Class–B stages are traditionally used in large-signal operation (above 10 mV) and in high-power operation (from 1 to 250 watts). Class–B stages have the advantage of greater operating efficiency in applications that involve substantial power. Because of low-level non-linearity in the transfer curve, class–B stages tend to develop crossover distortion unless an appreciable amount of negative feedback is em-ployed. Class–AB operation minimizes crossover distortion at some sacrifice of efficiency. Class–D stages are more efficient than class–B stages because the pulse form of signal permits an amplifier transistor to be driven to a higher power level than in conventional operation.

CHART 3–1
Basic Classes of Amplifier Operation

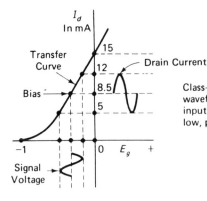

Class-A amplification provides an output waveform that is a precise replica of the input waveform. Amplifier efficiency is low, particularly during idling periods.

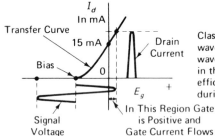

Class-B amplification provides an output waveform that is a half cycle of the input waveform; these half cycles are compressed in the low-current region. Amplifier efficiency is fairly good, particularly during idling periods.

(JFET Stage)

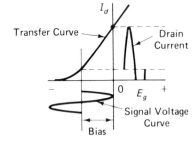

Class-AB amplification provides an output waveform that is more than a half cycle, but less than a full cycle of the input waveform. Amplifier efficiency is intermediate to that of Class-A and Class-B amplification.

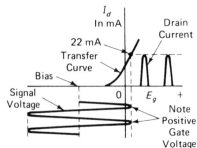

Class-C amplification provides an output waveform that is less than a half cycle of the input waveform. In audio technology, Class-C amplification represents amplifier malfunction. Amplifier efficiency is higher than for Class-B operation.

(JFET Stage)

CHART 3–1 *Continued*

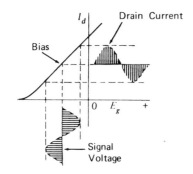

Class - D amplification provides an output that is a precise replica of the input waveform. The input signal is changed into a pulse form, amplified, and then integrated to restore the original waveform. Amplifier efficiency is higher than for Class - B operation.

Because of its bilevel mode of signal processing, class–G amplification (Figure 3–3) is also more efficient than class–B amplification. PWM amplification (Figure 3–4) is a variant of class–D operation and also provides high efficiency.

Advanced stereo systems occasionally employ bi-amp arrangements, as exemplified in Figure 3–5. Bass frequencies are amplified separately from treble frequencies in the output stage, because the low-frequency amplifier is preceded by a low-pass filter, and the high-frequency amplifier is preceded by a high-pass filter. This design permits each output amplifier to operate at half the bandwidth that would

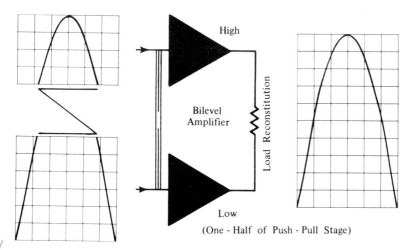

Figure 3–3 A class-G amplifier processes the high-level portion of a signal separately from its low-level portion.

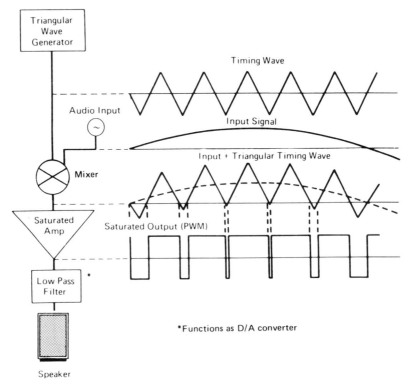

√ **Figure 3–4** PWM amplification is a high-efficiency pulse-type of transistor operation.

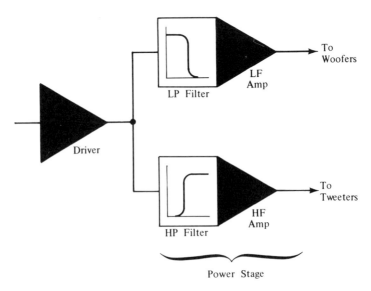

√ **Figure 3–5** Basic bi-amp arrangement.

otherwise be required. In turn, design problems are eased, and somewhat higher performance can be achieved. In particular, negative-feedback problems at the upper and lower ends of the passbands are relaxed. Moreover, since the low-frequency components in an audio signal contain more power than the high-frequency components, less reserve power capability is required in design of the high-frequency amplifier, and better system efficiency can be realized. Audio perfectionists tend to prefer bi-amp arrangements to the more conventional output-amplifier arrangements.

Extended High Frequency Response

Conventional high-fidelity amplifiers have frequency response up to 20 kHz or greater. Human hearing is limited to 15 kHz or less, but an exceptional listener may perceive frequencies as high as 17 kHz. A speaker system with a super tweeter can reproduce frequencies up to 40 kHz. Advanced stereo systems occasionally employ amplifiers with extended high-frequency response up to 250 kHz (see Figure 3–6).

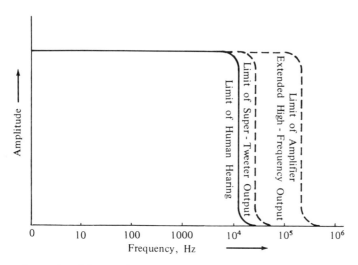

Figure 3–6 Amplifiers with extended high-frequency response ranges up to 250 kHz.

Conventional tape recorders reproduce frequencies up to approximately 15 kHz; but, a high-performance tape recorder can reproduce frequencies up to 20 kHz, or slightly higher. A high-performance phono cartridge can reproduce frequencies somewhat higher than 20 kHz; however, recordings may not provide high-frequency output beyond 15 khz. FM broadcasts provide high-frequency response up to 20 kHz. Most AM broadcasts are limited to 5 kHz, but an occasional high-fidelity AM station provides high-frequency response up to 10 kHz. High-performance microphones used with advanced stereo systems have high-frequency output up to 15 kHz.

Frequency Response and Power Bandwidth

Output amplifiers for advanced stereo systems are generally rated for bandwidth (within ± 1 dB) at maximum rated power output with a reference distortion value at 1 kHz, such as 1 percent (see Figure 3–7).

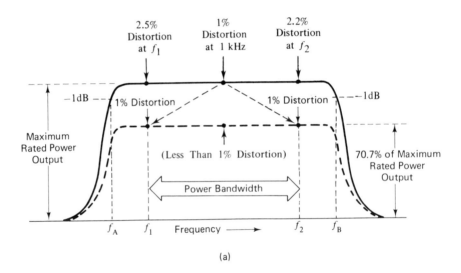

(a)

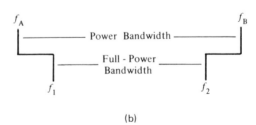

(b)

Figure 3–7 Example of amplifier power bandwidth characteristics referenced to 1 percent distortion at maximum rated power output at 1 kHz. **(a)** Typical full-power bandwidth/power bandwidth relation; **(b)** another example of full-power bandwidth/power bandwidth relation.

This rating, from f_A to f_B, is termed the full-power bandwidth of the amplifier. Advanced stereo amplifiers are generally rated for power bandwidth as well. As exemplified in the diagram, power bandwidth is specified at -3 dB of maximum rated power output and with the reference distortion value (1 percent in this instance). Thus, the power bandwidth for this particular amplifier extends from f_1 to f_2. Note that the -3 dB level is also called the half-power level, or the 70.7 percent voltage level. As a general rule of thumb, the power-bandwidth rating of an amplifier is less than its full-power bandwidth rating. However, some exceptions occur in designs of advanced stereo power amplifiers that have a power-bandwidth rating greater than their full-power band-width rating.

INTEGRATED AMPLIFIERS AND SEPARATES

Advanced stereo systems may employ either integrated amplifiers or separate preamplifiers and power amplifiers. When substantial audio power output is contemplated (for example, more than 100 watts per channel), the comparative bulk of the output sections often influences the audiophile's decision to choose separates. In turn, the power amplifiers can be installed apart from the comparatively compact control center with its preamplifiers. A super-high-power amplifier with high performance characteristics is illustrated in Figure 3–8. Observe that no operating controls are provided. An interconnection diagram for the amplifier is shown in Figure 3–9. Note the comparative sizes of the preamplifier and the super-high-power amplifier. In many listening areas, the super-high-power amplifier might be installed away from the preamplifier. Specifications for the super-high-power amplifier are given in Figure 3–10. Note that complete overload and speaker protec-tion is provided; this is an essential consideration in an advanced stereo system that operates at high power-output levels.

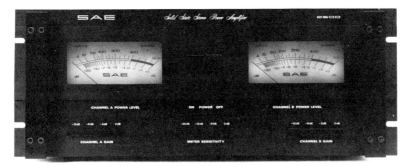

Figure 3–8   Appearance of an advanced audio super-high-power amplifier. (*Courtesy, Scientific Audio Electronics, Inc.*)

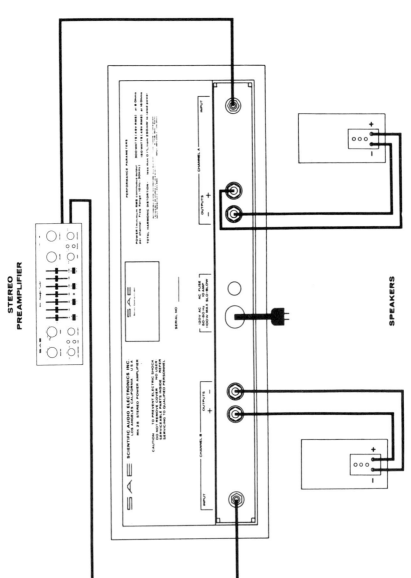

Figure 3–9 Interconnection diagram for the SAE super-high-power amplifier.

RMS (min) continuous power output per channel 20 Hz to 20 kHz (both channels driven) at 8 Ω. 300 W at 0.05% total harmonic distortion
THD (total harmonic distortion) from 20 Hz to 20 kHz at 250 mW to rated power at 8 Ω. 0.05% max.
IM (intermodulation distortion) from 250 mW to rated power at 8 Ω with any two mixed frequencies between 10 Hz and 30 kHz at 4:1 voltage ratio. 0.05% max.
Frequency response at rated power. ±0.25 dB, 10 Hz to 30 kHz
Noise. Greater than 100 dB below rated power
Transient response of any square wave. 2.5 μs rise and fall time
Slew rate. 40 V/μs
Stability. Unconditionally stable with any type of load or no load including full-range electrostatic loudspeakers
Damping factor. 150 min (100 Hz)
Input sensitivity. 1.5 volts rms for rated output at 8 Ω
Input impedance. 50 kΩ
Semiconductor complement. 46 transistors, 49 diodes
Overload protection. 1. Low-impedance electronic-sensing circuit limits with output current below 2 Ω without limiting with 4 Ω or higher (or reactive loads).
2. Thermal sensing of inadequate ventilation.
3. Internal B+, B− supply fuses.
Loudspeaker protection. Relay circuit protects loudspeakers from low-frequency oscillations and plus or minus dc output. Five-second turn on/off delay eliminates on/off disturbances.
Power requirements. 110–125 V 50 Hz/60 Hz, 100 W at idling, to 1100 W at rated output.

Figure 3–10   Specifications for the SAE super-high-power amplifier.

Often separates are chosen also for biamplifier operation in advanced stereo systems. In Figure 3–11, a typical biamp system employs a bass power amplifier and a treble power amplifier with an electronic crossover unit that provides high-pass, low-pass, band-pass, band-reject, and adjustable corner-frequency characteristics. This electronic crossover unit is driven by a separate preamplifier. A conventional high-performance separates arrangement with microphones, turntable, and tuner is shown in Figure 3–12. As in the case of the assembly depicted in Figure 3–9, the power amplifier is much bulkier than the preamplifier; consequently, the listener may prefer to install the power amplifier apart from the preamplifier. For aficionados who prefer separates in housings with the same size and similar styling, arrangements such as the type pictured in Figure 3–13 are available.

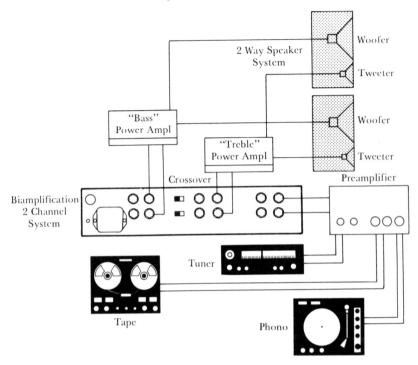

Frequency Response: 18 Hz to 38 kHz ±0.5-dB into 600-ohm load.
Inputs: Bridging input, 20 K ohm balanced or 10 K ohm unbalanced, and 1 M ohm unbalanced; both using ¼ in. standard phone jack.
Output: 10 volts maximum before overload; 2.5 volts rated.
Gain: 0 to 15.5 dB from balanced/unbalanced input.
Hum and Noise: More than 100-dB below rated output, with 0-dB gain, over entire audio spectrum from 20 Hz to 20 kHz.
IM Distortion: Less than 0.01% at rated output.
Filter Characteristics: Separate 18-dB Butterworth highpass and lowpass, with adjustable corner frequencies. Can be internally cascaded to form band pass and band reject filters.
Controls: (front panel) Range and vernier controls for corner frequencies (high and low pass), power on/off switch. (rear panel): Screwdriver-adjustable input attenuators for each channel.

(a)

√ **Figure 3–11**   Interconnection diagram and specifications for the Crown International VFX-2 power biamplifier.

An elaborate top-performance separates system with a flexible preamplifier unit is illustrated in Figure 3–14. It comprises an advanced FM/AM tuner and a pair of 200–watt super-power amplifiers. The preamplifier includes a digital tuner and optional plug-in modules for various sophisticated functions. Thus, Dolby FM may be added to the basic assembly; SQ or CD–4 quadraphonic modules may be

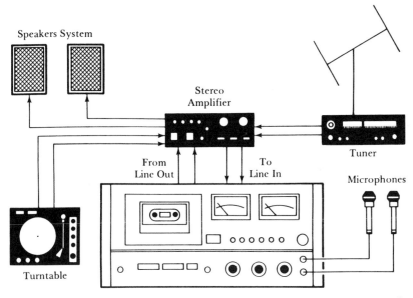

Figure 3–12   Interconnection diagram for a separates arrangement with microphones, turntable, and tuner. *(Courtesy, Pioneer)*

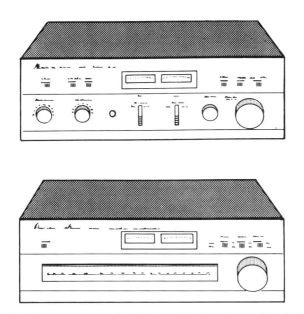

Figure 3–13   Separates may be designed in housings with similar styling and size.

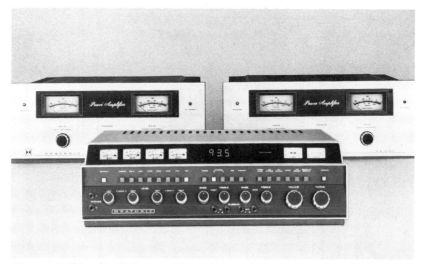

Figure 3–14   An elaborate top-performance separates system with a flexible preamplifier unit. *(Courtesy, Heath Co.)*

employed, or an alternative design of amplifier may be used. Two tuning meters are provided; one meter indicates signal strength for AM or FM reception, and the other functions as an FM tuning meter. Peak-responding meters are provided for the L and R channels for monitoring inputs to the power amplifiers. Frequency response of the power amplifiers is within 1 dB from 7 Hz to 50 kHz, with less than 0.1 percent distortion at maximum rated power output.

ADVANCED AMPLIFIER BASICS

Distortion is minimized in high-fidelity amplifiers by means of negative-feedback action, as exemplified in Figure 3–15. Negative feedback also changes the input and output impedances of an amplifier stage, as explained in Table 3–2. The following two basic guidelines are often helpful in the analysis of negative-feedback circuitry:

1. When a device employs a large amount of current feedback (such as a high-valued emitter resistor), the input port of the stage is almost an open-circuit.

2. When a device employs a large amount of voltage feedback (such as a low-valued resistor between collector and base), the input port of the amplifier is a virtual ground.

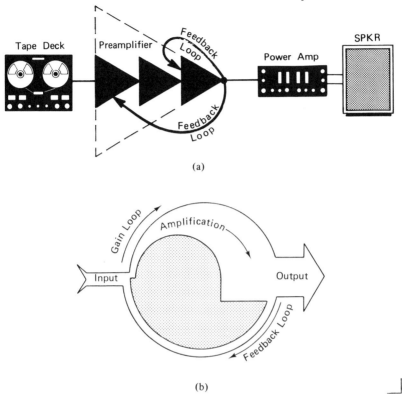

(a)

(b)

Figure 3–15 Negative feedback. **(a)** Feedback loops minimize distortion; **(b)** basic principle of negative feedback.

Advanced stereo amplifiers may have quasi-complementary configuration, as developed in Chart 3–2. This arrangement uses negative feedback to minimize distortion; it also exploits positive feedback (bootstrapping) to equalize the outputs from the two sides of the complementary network. In Figure 3–16, the input voltage E_{in} appears in the same phase and at slightly reduced amplitude across R_E. This in-phase voltage is fed back via C and R1 to the base of Q. This fraction of the drop across R_E adds regeneratively to E_{in} and thereby increases the input impedance to the stage. This is a form of positive feedback. However, self-oscillation cannot occur in a bootstrapped circuit because the signal voltage across the emitter resistor is always less than the source input voltage. Note in passing that the current gain of the stage is somewhat reduced owing to the fraction of the output signal that is diverted through the bootstrap capacitor.

TABLE 3-2

INPUT AND OUTPUT IMPEDANCES OF AMPLIFIER STAGES WITH NEGATIVE FEEDBACK

Input Impedance (Resistance) of Common-Collector Stage

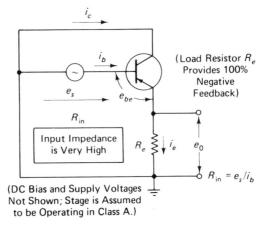

(DC Bias and Supply Voltages
Not Shown; Stage is Assumed
to be Operating in Class A.)

The AC Input Impedance, or Resistance R_{in} is Very High in the Common-Collector Configuration Because of the Negative-Feedback Action in the Emitter Circuit.

Consider that $R_e = 0$; then $e_o = 0$, $e_{be} = e_s$, and i_b has its maximum value. Next, if R_e has a substantial resistance value, e_o will be almost as large as e_s, and e_{be} will be small; e_{be} is equal to the difference between e_s and e_o. When e_{be} becomes smaller, i_b also becomes smaller. Consequently, R_{in} becomes large; R_{in} is equal to e_s/i_b, in accordance with Ohm's law.

Input Impedance (Resistance) of Common-Emitter Stage with Emitter Current Feedback

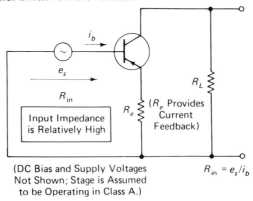

(DC Bias and Supply Voltages $R_{in} = e_s/i_b$
Not Shown; Stage is Assumed
to be Operating in Class A.)

The AC Input Impedance, or Resistance R_{in} is Rather High in the Common-Emitter Configuration When There is Substantial Negative-Feedback Action in the Emitter Circuit.

A common-emitter configuration with emitter feedback has an input impedance (resistance) that is higher than that of a CE stage without emitter feedback, but lower than that of the common-collector configuration. This relation follows from the condition of less than 100 percent negative feedback in the CE configuration with an unbypassed emitter resistor.

TABLE 3–2 *Continued*

Output Impedance (Resistance) of Common-Emitter Stage with Emitter Current Feedback

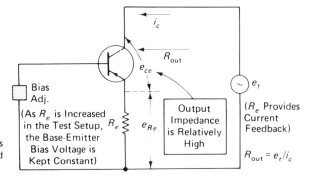

(DC Bias and Supply Voltages Not Shown; Stage is Assumed to be Operating in Class A.)

Bias Adj.

(As R_e is Increased in the Test Setup, the Base-Emitter Bias Voltage is Kept Constant)

Output Impedance is Relatively High

(R_e Provides Current Feedback)

$R_{out} = e_t/i_c$

The AC Output Impedance, or Resistance R_{out} is Rather High in the Common-Emitter Configuration When There is Substantial Negative-Feedback Action in the Emitter Circuit.

Consider that $R_e = 0$; then e_{ce} will be equal to the test voltage e_t. Next, if R_e has a high resistance value, e_{Re} will be high; in turn, e_{ce} becomes smaller because e_{ce} is equal to the difference between e_t and e_{Re}.

Consequently, i_c becomes smaller and R_{out} becomes higher because R_{out} is equal to e_t/i_c in accordance with Ohm's law.

Input Impedance (Resistance) of Common-Emitter Stage with Voltage Feedback

(R_f Provides Voltage Feedback)

Input Impedance is Relatively Low

$R_{in} \doteq e_s/i_{in}$

(DC Bias and Supply Voltages Not Shown; Stage is Assumed to be Operating in Class A.)

The AC Input Impedance, or Resistance R_{in} is Rather Low in the Common-Emitter Configuration When Voltage Feedback is Used.

Consider that $R_f = \infty$; then the input resistance is equal to e_s/i_b.

Next, if R_f has a moderate value, the total input current i_{in} becomes greater; i_{in} is then equal to $i_b + i_c$, where i_c is the "cancellation current" due to mixture of e_s with the out-of-phase feedback voltage.

Since i_{in} is greater when negative feedback is used, the input resistance R_{in} decreases, because $R_{in} = e_s/i_{in}$, according to Ohm's law.

TABLE 3-2 *Continued*

Output Impedance (Resistance) of Common-Emitter Stage with Voltage Feedback

(R_f Provides Voltage Feedback)

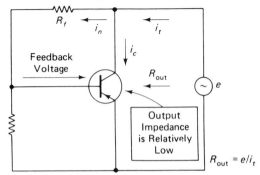

$$R_{out} = e/i_t$$

(DC Bias and Supply Voltages
Not Shown; Stage is Assumed
to be Operating in Class A.)

The AC Output Impedance, or Resistance R_{out} is Rather Low in the Common-Emitter Configuration When Voltage Feedback is Used.

Consider that $R_f = \infty$; then the output resistance is equal to e/i_c.

Next, if R_f has a moderate value, the total current i_t becomes greater; i_t is then equal to $i_c + i_n$, where i_n is the "neutralization" or cancellation current due to mixture of e with the amplified feedback voltage that appears at the collector.

Since i_t is greater when negative feedback is used, the output resistance R_{out} decreases, because $R_{out} = e/i_t$, in accordance with Ohm's law.

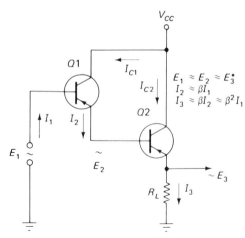

$$E_1 \approx E_2 \approx E_3^*$$
$$I_2 \approx \beta I_1$$
$$I_3 \approx \beta I_2 \approx \beta^2 I_1$$

$* \approx$ Means "Is Approximately
Equal to".

TABLE 3–2 *Continued*

Input Impedance (Resistance) of Darlington Stage

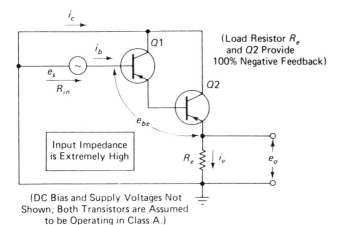

(Load Resistor R_e and Q2 Provide 100% Negative Feedback)

(DC Bias and Supply Voltages Not Shown; Both Transistors are Assumed to be Operating in Class A.)

The AC Input Impedance, or Resistance R_{in} is Extremely High in the Darlington Configuration Because of the Circuit's Beta Multiplication.

Consider first that Q1 is omitted from the circuit, and that e_s is applied directly to the base of Q2. Under this condition of operation, a simple common-collector configuration results, and its input impedance is very high, as explained previously.

Next, with Q1 included in the circuit, the beta of Q1 becomes multiplied by the beta of Q2. Consequently, i_e becomes larger, and e_{be} becomes smaller. When e_{be} becomes smaller, i_b also becomes smaller than in a common-collector configuration. In turn, R_{in} is larger; R_{in} is equal to e_s/i_b, in accordance with Ohm's law.

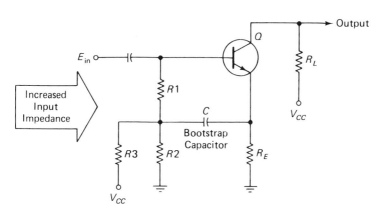

Figure 3–16   Basic example of a bootstrapped single-stage amplifier.

91

CHART 3-2

DEVELOPMENT OF THE QUASI-COMPLEMENTARY
SYMMETRY AMPLIFIER

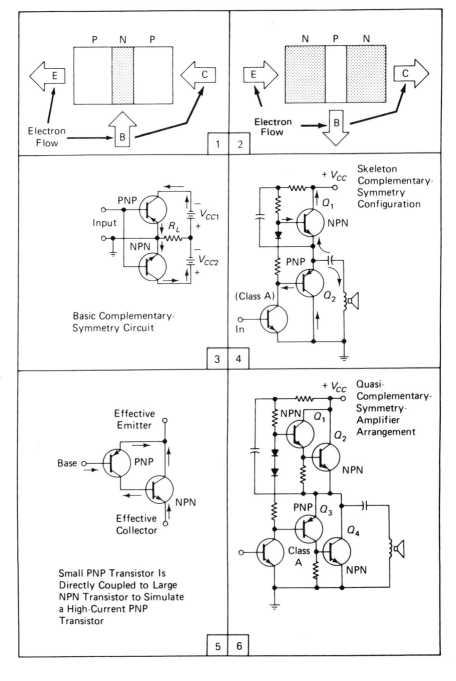

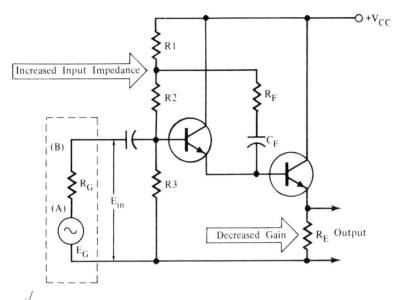

√ **Figure 3–17**   Example of a bootstrapped two-device stage.

Observe the example of a bootstrapped two-device stage shown in Figure 3–17. This is a Darlington connection with additional feedback components C_F and R_F. These components apply in-phase feedback voltage to the base of Q1 via R2. In turn, the signal voltage at (B) is effectively increased and becomes more nearly equal to the signal voltage at (A). Accordingly, the current through R_G is reduced, and the effective input impedance of Q1 is increased. On the other hand, the stage gain is reduced somewhat owing to the fraction of the output signal that is diverted through C_F. This bootstrap arrangement is widely employed in quasi-complementary amplifier configurations. The input impedance to the stage is adjusted to the optimum value by varying the R1/R2 ratio. Note that a Darlington amplifier is not a voltage amplifier; it operates as a current amplifier and its output voltage is slightly less than its input voltage. A Darlington amplifier is also a power amplifier; power output is proportional to current squared, as illustrated in Figure 3–18.

VERTICAL FIELD-EFFECT TRANSISTORS

Although most advanced stereo amplifiers have bipolar (conventional) transistors, unipolar (field effect) power-type transistors are becoming more common because of their superior transient response. Transient

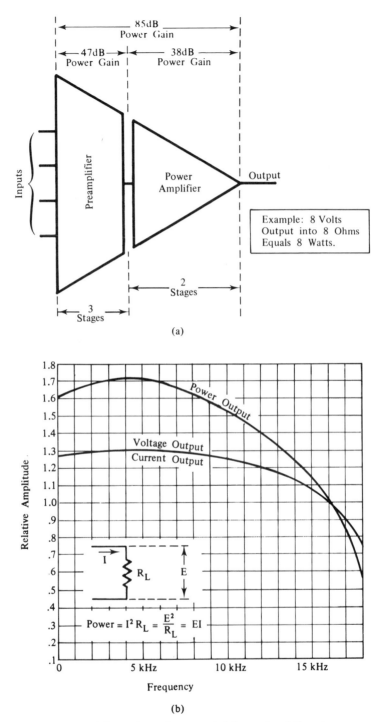

Figure 3–18 Amplifier gain and output. (**a**) Example of five-stage arrangement with 85 dB power gain; (**b**) corresponding relations of frequency response curves for voltage or current output, and for power output.

response is chiefly a function of charge carrier storage, which tends to slow down the slew rate of power-type bipolar transistors. Field-effect transistors are comparatively free from charge-storage delay because they are voltage-operated devices instead of current-operated devices. In turn, unipolar power-type transistors have better frequency response and lower inherent distortion than do bipolar transistors. Although negative feedback must be employed in power amplifiers that utilize field-effect transistors, less feedback is required to obtain a satisfactorily low distortion level. Field-effect power-type transistors also have better overload characteristics than do bipolar transistors. In other words, if the amplifier is accidentally overloaded by an unusually high-amplitude audio pulse, the resulting distortion is less objectionable than if the peak signal had been clipped by bipolar transistors.

Field-effect transistors are fabricated in planar form on a chip that is approximately 0.7 mm square, as depicted in Figure 3–19(a). On the other hand, a power-type junction field-effect transistor (JFET) or vertical FET (V–FET) uses a chip that is about 3 mm square. As shown in Figure 3–19(b), approximately 1500 tiny source areas are interconnected by metallic jumpers. In turn, the current flow divides into many paths from source to drain, and the device has comparatively high power-handling capability. Figure 3–20 shows the basic configuration for a V–FET amplifier. This arrangement provides 150 watts rms output with less than 0.1 percent distortion over a frequency range of 20 Hz to 20 kHz. Approximately 40 dB of negative feedback is provided with a damping factor of 100 at frequencies below 1 kHz. Differential input is used, followed by cascode circuitry that uses a neutralized grounded-source input stage working into a grounded-gate output stage. This mode of operation provides high gain, high input impedance, and low noise. The second-generation V-MOSFET (VMOS)

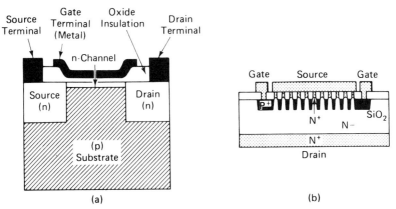

Figure 3–19   Field-effect transistor structure. **(a)** Conventional construction; **(b)** power-type vertical (V-FET) construction.

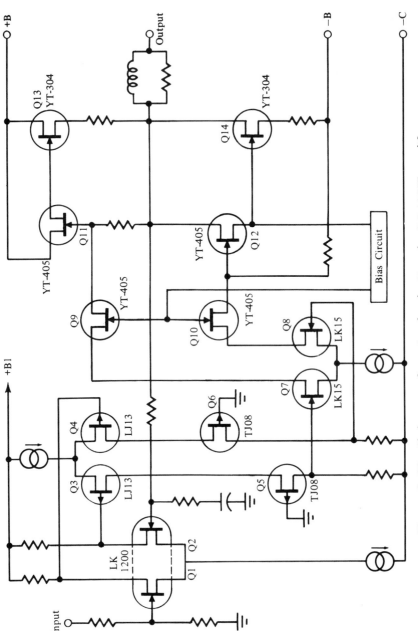

Figure 3–20 Skeleton configuration for the Yamaha B-1 V-FET amplifier.

power-type transistor produces up to 80 watts of output power at low frequencies (audio frequencies) with only one microwatt of input power. The VMOS transistor is well adapted to Class-D amplification because of its fast switching capability and its zero charge-carrier storage time.

DYNAMIC HEADROOM

In the past, amplifiers have been rated both for sine-wave (rms) power-output capability, and for music-power capability. The latter rating is based on the ability of a power amplifier to provide short "pulses" of output voltage that have comparatively high amplitudes. In other words, the majority of musical and vocal signals are characterized by narrow peak waveforms, instead of sine waveforms. A power amplifier's potential for reproducing high-amplitude narrow pulses without objectionable distortion depends chiefly upon the values of the filter capacitors in its power-supply section. In other words, a power-output level that exceeds the maximum rms rating of an amplifier is accompanied by high current drain from the power supply, and soon causes decreased output voltage owing to the regulation factor of the power supply. However, the peak-power rating for pulse output from a typical power amplifier may be 50 percent greater than its maximum rms-power rating. Because pulse signals do not precisely correspond to the peak waveforms in vocal and musical passages, this music-power rating has been an equivocal rating factor.

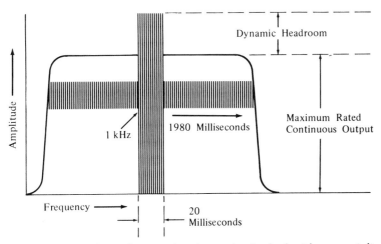

Figure 3–21 Amplifier dynamic headroom is checked with a specialized IHF tone-burst signal.

The recent trend to replace music-power ratings with a power-output value termed the dynamic headroom rating of the amplifier is depicted in Figure 3–21. Dynamic headroom is measured with a specialized Institute of High Fidelity (IHF) tone-burst signal. This signal consists of a 1–kHz sine wave that drives the amplifier at a comparatively low level for 1980 milliseconds, and then suddenly drives the amplifier to its maximum output capability for 20 milliseconds. There is a disparity of 10 dB between the low-level component and the high-level component of the tone-burst signal. An oscilloscope is used to monitor output from the amplifier under test. The signal amplitude is gradually increased until the 20 millisecond burst starts to display distortion. This amplitude defines the dynamic headroom rating of the amplifier.

POWER BOOSTER FOR AUTOMOBILE INSTALLATION

Higher power operation is becoming more common in advanced stereo systems for motor vehicles. As an illustration, the unit pictured in Figure 3–22 provides an audio power output of 60 watts for the channel in which it is used. A pair of these boosters employed in the L and R channels will provide a total audio output of 120 watts. A seven-band frequency equalizer is included in the unit to assist in compensation for acoustic anomalies in the automotive aural environment. The unit is conveniently mounted under the dash of the automobile.

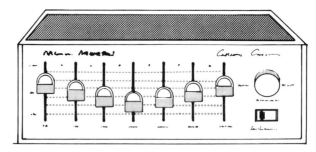

Figure 3–22 Under-dash, 60-watt power booster with seven-band equalizer. *(Courtesy, Marantz)*

4

High-Performance Integrated Receivers

GENERAL CONSIDERATIONS

An integrated receiver such as the one illustrated in Figure 4–1 contains an FM/AM tuner, preamplifier, and power amplifier. Advanced stereo integrated receivers are characterized by a high value of signal-to-noise ratio such as 50 dB or more, a low percentage of distortion, a high value of selectivity such as 50 dB or greater, a high value of AM rejection by the stereo section such as 50 dB, an FM image rejection capability of 80 dB or more, and a stereo separation value in the range from 20 to 30 dB. In addition, integrated receivers are rated for "headroom" as previously noted, for different deemphasis characteristics, for phono-preamp sensitivity levels, rumble and scratch filter characteristic, for capture ratio, and for tuner sensitivity. A basic plan for a typical integrated receiver is shown in Figure 4–2. The tuning meter is a guide to optimum FM response, and the signal-strength meter indicates comparative levels of AM signals.

The "capture effect" is graphically portrayed in Figure 4–3. This is a circuit action whereby an FM discriminator automatically suppresses the weaker of two interfering signals and passes on the stronger signal. Thereby, co-channel interference is eliminated in favor of the stronger signal, provided that its amplitude exceeds the capture-ratio threshold of the discriminator. In an advanced stereo integrated receiver, a typical capture-ratio rating is 1.5 dB. This means that if the amplitudes of two interfering signals differ by at least 1.5 dB, the weaker signal will be completely suppressed by the discriminator network. An advanced stereo integrated receiver is also designed to

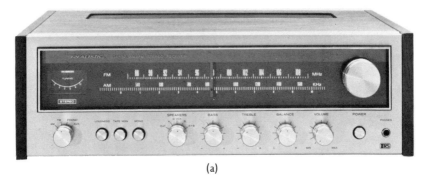

(a)

(b)

Figure 4–1   Integrated receivers. **(a)** A high-performance AM/FM stereo receiver *(Courtesy, Radio Shack);* **(b)** a sophisticated AM/FM stereo receiver *(Courtesy, Lafayette Electronics)*

provide a high degree of image rejection. Figure 4–4 illustrates the frequency relations between a desired signal, local-oscillator frequency, and an image-interference signal. When the FM receiver is tuned to the desired signal frequency (100 MHz), the local oscillator operates at 110.7 MHz. A very strong interfering signal with a frequency of 121.4 MHz (image frequency) could penetrate the front-end circuitry and cause audible interference. This type of interference is most common in the vicinity of airports. As noted above, an image-rejection ratio of 80 dB is typical of an advanced stereo integrated receiver.

Separation denotes the extent to which L and R stereo channels are free from crosstalk. In Figure 4–5, zero dB separation denotes that there are equal amounts of L and R signals in each channel. In this situation, stereo reproduction is entirely lacking, and reproduction is monophonic. Note that 1 dB separation is barely discernible; reproduction is essentially monophonic. A separation of 20 dB provides

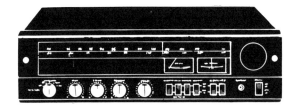

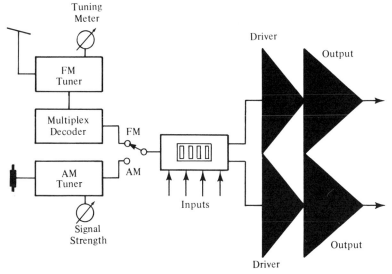

Figure 4–2 Basic plan of an integrated receiver.

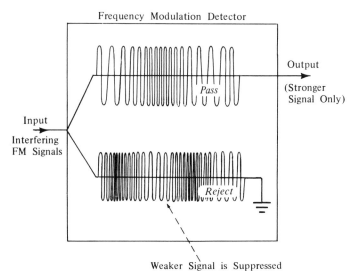

Figure 4–3 Visualization of the "capture effect" provided by an FM discriminator.

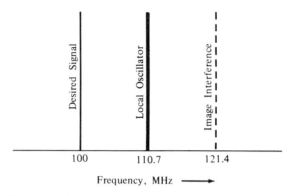

Figure 4–4 Example of image-frequency rejection.

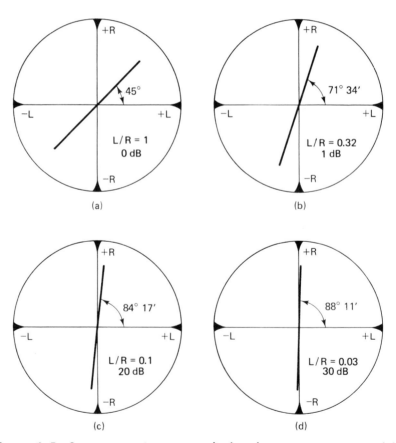

Figure 4–5 Stereo separation patterns displayed on vectorscope screen. **(a)** Zero dB separation; **(b)** 1 dB separation; **(c)** 20 dB separation; **(d)** 30 dB separation.

satisfactory separation of the L and R channel signals. Advanced stereo receivers are often capable of 30 dB separation. Note in passing that 30 dB separation could become seriously degraded to a value of 5 dB, for example, if the L and R speakers were installed too near each other. In other words, the separation rating of an integrated receiver applies to its electrical channels only, and has no relation to the presence or absence of acoustic separation. Acoustic separation is solely a characteristic of the listening area, as previously noted in Chapter 2.

SELECTIVITY CHARACTERISTICS

Selectivity in an advanced stereo tuner is based on the characteristics of resonant configurations in comparatively sophisticated designs. As shown in Figure 4–6, FM broadcast channel allocations are established from 88 to 108 MHz. Each channel encompasses a "swing" of ± 75 khz, or a total excursion of 150 kHz. A guard band of 50 kHz is provided between adjacent channels. If ideal selectivity could be realized in an FM tuner, its frequency response curve would have a rectangular contour. In practice, ideal selectivity can only be approximated. Consequently, advanced stereo tuners achieve selectivity factors in the order of 50 dB and image rejection ratios in the order of 80 dB. This degree of selectivity is entirely satisfactory for operation in all but extremely difficult situations. For example, the most sophisticated FM tuner available is not likely to reject image interference completely if it is operated within several hundred feet of an airport.

Tapped inductors are widely used in resonant-circuit design for FM tuners. Figure 4–7 shows that the resonant frequency and the unloaded Q value of a parallel-resonant circuit remain the same whether the signal is injected across points 1 and 3, or across points 2 and 3. The input impedance between points 2 and 3, however, is much smaller than the input impedance between points 1 and 3, depending on the square of the turns ratio between points 1 and 3, and between points 2 and 3. For example, if point 2 is a center-tap, the input impedance between points 1 and 3 is four times the input impedance between points 2 and 3. In other words, if point 2 were one-third above point 3, the input impedance between points 1 and 3 would be nine times the input impedance between points 2 and 3. To continue with the same example, the input impedance between points 1 and 3 would be 2¼ times the input impedance between points 2 and 3.

Consider next the tapped arrangement depicted in Figure 4–7(c). The resonant frequency and the unloaded Q value remain the same, whether the signal is injected across points 1 and 3, or across points 2 and 3. However, the input impedance between points 2 and 3 is

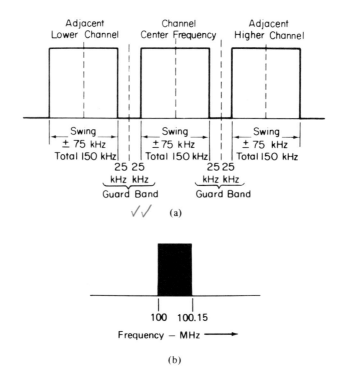

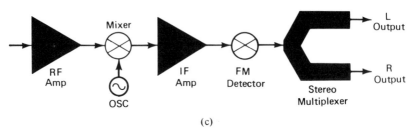

Figure 4–6 Frequency relations in FM tuner operation. **(a)** FM broadcast channel allocations; **(b)** ideal FM tuner frequency response; **(c)** functional diagram of an FM tuner; **(d)** view of a rack-mount advanced stereo tuner (*Courtesy, Heath Co.*);

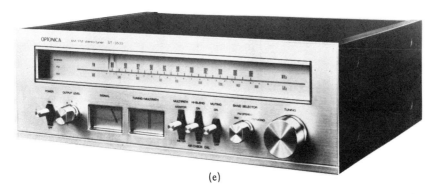

(e)

Figure 4–6 *Continued* **(e)** an advanced AM/FM tuner with multigraph monitor *(Courtesy, Optonica).*

smaller than the input impedance between points 1 and 3. For example, if C1 and C2 have equal values, the input impedance between points 1 and 3 is four times the input impedance between points 2 and 3. This difference is evident, because if point 2 were connected to a center tap on L, circuit action would remain unchanged in that a balanced bridge configuration results. The general C versus Z relation is shown in Figure 4–8.

In many designs, a tuned resonant circuit (capacitor C_p and inductor L_p) in the primary of a tuned transformer is coupled to the non-resonant secondary of the transformer, as depicted in Figure 4–7(d). In this case, if N1 represents the number of turns in the primary winding, and N2 represents the number of turns in the secondary winding, then the turns ratio m of primary to secondary under matched conditions is equal to the square root of the ratio of the impedances that are to be matched. If there is capacitance in the secondary circuits C_s, it is reflected (or referred) to the primary circuit by transformer action as a capacitance C_{sp}. Its effective value in the primary circuit is equal to the ratio of capacitance C_s to the square of the turns ratio.

The output impedance of a transistor can be considered a resistance r_o parallel to a capacitance C_o, as shown in Figure 4–9(a). Similarly, the input impedance of a transistor can be considered a resistance r_i parallel to a capacitance C_i, as indicated in the diagram. In most situations, the output capacitance and the input capacitance are considered to be part of the coupling network, as represented in Figure 4–9(b). If an untuned primary and a tuned secondary are used, as in Figure 4–9(c) and (d), the frequency response characteristic is far from ideal (refer to Figure 4–6(b)). This departure from ideal

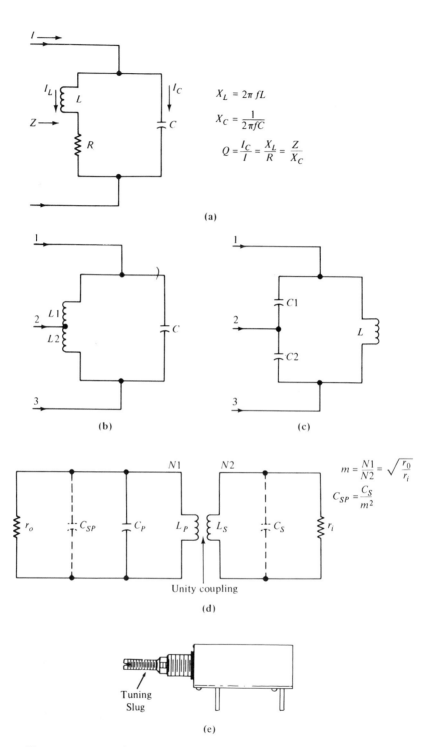

$$X_L = 2\pi f L$$

$$X_C = \frac{1}{2\pi f C}$$

$$Q = \frac{I_C}{I} = \frac{X_L}{R} = \frac{Z}{X_C}$$

(a)

(b)

(c)

$$m = \frac{N1}{N2} = \sqrt{\frac{r_0}{r_i}}$$

$$C_{SP} = \frac{C_S}{m^2}$$

Unity coupling

(d)

Tuning
Slug

(e)

Figure 4–7 Fundamental relations in basic parallel-resonant circuits.

selectivity can be reduced to some extent by cascading tuned stages; for example, the result of cascading up to four tuned stages is shown in Figure 4–8(b). However, this approach obviously only approximates an ideal frequency characteristic. Therefore, more sophisticated design measures are required in advanced stereo tuners.

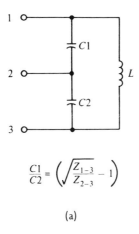

$$\frac{C1}{C2} = \left(\sqrt{\frac{Z_{1-3}}{Z_{2-3}}} - 1 \right)$$

(a)

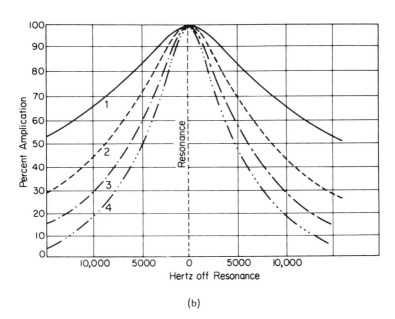

(b)

Figure 4–8   Basic resonant-circuit relations. **(a)** Impedance ratios and capacitance ratios; **(b)** elementary frequency response curves.

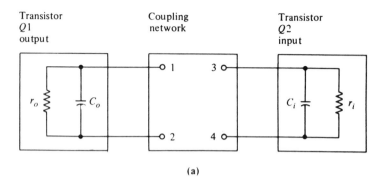

(a)

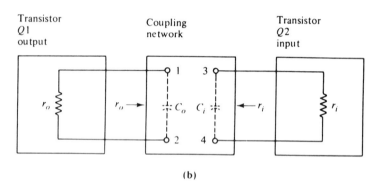

(b)

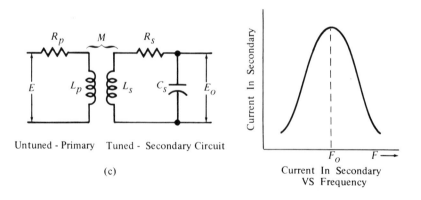

Untuned - Primary   Tuned - Secondary Circuit

(c)

Current In Secondary
VS Frequency

(d)

Figure 4–9 Equivalent output and input circuits with a coupling network, and example of frequency response.

TRANSFORMER COUPLING WITH TUNED PRIMARY

A generalized representation of a single-tuned circuit used as coupling network, and utilizing inductive coupling, is shown in Figure 4–10(a). In this circuit, capacitance C_T represents the output capacitance of transistor Q1 and the input capacitance of transistor Q2 referred to the primary of transformer T1. A match of the output impedance of Q1 to the input impedance of Q2 is obtained by means of a suitable turns ratio for T1. If L_p represents the inductive reactance between terminals 1 and 2 of T1, analysis will show that the frequency response of the configuration is the same as that for an elementary LC tuned circuit. Suppose that the arrangement in Figure 4–10(a) is used for an intermediate-frequency (IF) amplifier with two transistors operated in the common-emitter (CE) mode. The frequency response curve will then have the contour shown in (a) of Figure 4–8(b). To obtain proper bandwidth, a tapped primary must be employed, as depicted in Figure 4–10(b).

Inductive coupling from the output of one transistor to the input of a second transistor can be accomplished with a tuned autotransformer, as shown in Figure 4–11(a). The circuit action is almost the

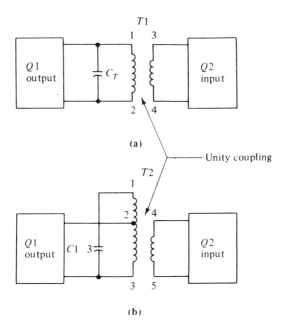

(a)

(b)

Figure 4–10   Interstage transformer coupling with tuned primary, and series-resonant circuit properties.

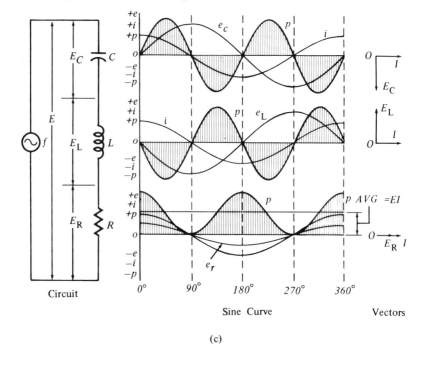

Circuit   Sine Curve   Vectors

(c)

Figure 4–10 *Continued*

same as for the arrangement in Figure 4–10, except that DC isolation is not provided between transistors Q1 and Q2. Capacitance C_T in Figure 4–11(a) includes the output capacitance of Q1 and the input capacitance of Q2 referred to the primary section. If L_{1-3} designates the inductance between terminals 1 and 3 of the autotransformer, then the resonant frequency of the arrangement can be calculated in the usual manner, and the frequency response curve will be the same as in Figure 4–9(d). The tap at terminal 2 on the autotransformer provides an impedance match between Q1 and Q2, as explained previously. In some applications, the selectivity may be inadequate; in such a case, the primary inductance should be increased. This difficulty may be circumvented, however, by using the alternative autotransformer arrangement depicted in Figure 4–11(b). To maintain a particular resonant frequency, C_{1-4} must be reduced by the same factor so that the product of L_{1-3} and C_T in (a) is equal to the product of L_{1-4} and C_{1-4} in (b). Impedance matching is maintained, provided that inductances L_{2-4} and L_{3-4} of transformer T2 equal the inductances L_{1-3} and L_{2-3}, respectively, of transformer T1.

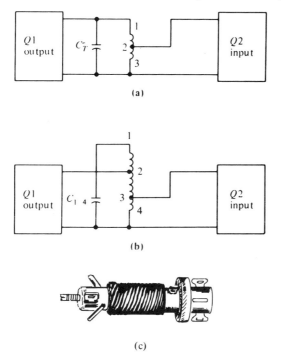

(a)

(b)

(c)

Figure 4–11 Interstage coupling network utilizing autotransformer circuitry with a tuned primary winding.

CAPACITANCE COUPLING

When a transformer coupling problem arises in an advanced stereo tuner design that limits the secondary to a comparatively small number of turns, unity coupling, or even tight coupling, between primary and secondary may be impractical. The input impedance to the transistor is less than 75 ohms in this situation. To circumvent this difficulty, capacitance coupling is often used, as in Figure 4–12. Impedance matching of Q1's output impedance to Q2's input impedance is obtained by selecting the suitable ratio of C_1 to C_2. Capacitance C_2 is ordinarily much larger than C_1. Their reactance values bear the opposite relationship to their capacitance values. If inductance L_1 is too small to provide the required selectivity, the configuration shown in Figure 4–12(b) may be used. With a tapped coil, the inductance may be increased by a chosen factor, and the total capacitance C_T reduced by the same factor in order to maintain the same resonant frequency. Impedance matching is obtained by making the inductance between

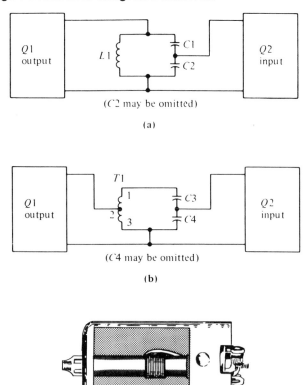

Figure 4–12 Interstage capacitance coupling with a split capacitor.

terminals 2 and 3 equal to L_1; the ratio of C_3 to C_4 must be equal to the ratio of C_1 to C_2.

INTERSTAGE COUPLING WITH DOUBLE-TUNED NETWORKS

Advanced stereo tuners generally employ interstage coupling with double-tuned networks, particularly in the intermediate-frequency (IF) section. Advantages of double-tuned interstage coupling networks, as shown in Figure 4–13, include a flatter frequency response within the pass band, a sharper drop in response immediately adjacent to the ends of the pass band, and comparatively high attenuation of frequencies not in the pass band. Thus, a better approach to optimum selectivity characteristics (Figure 4–6(b)) is realized. Observe the frequency

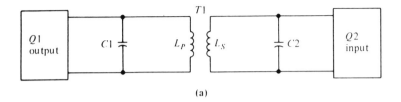

(a)

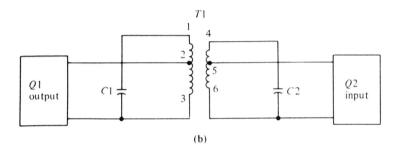

(b)

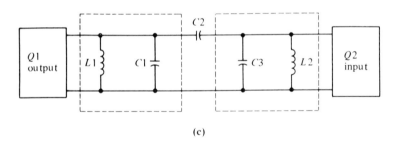

(c)

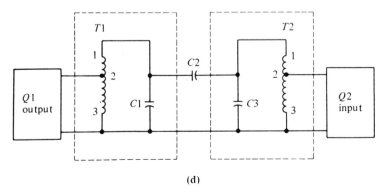

(d)

Figure 4–13   Basic double–tuned interstage coupling networks.

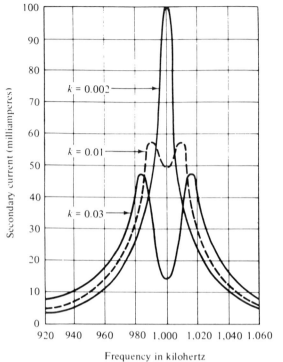

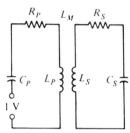

$L_P = L_S = 0.15$ mH
$C_P = C_S = 169$ pF
$R_P = R_S$
$Q_P = Q_S = 100$

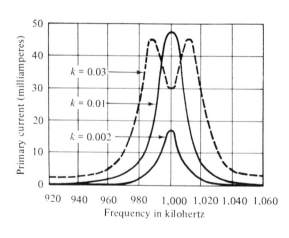

Figure 4–14   Primary and secondary frequency characteristics for a double–tuned transformer with three coefficients of coupling.

response curves in Figure 4–14 for a double-tuned transformer with three coefficients of coupling. Inductively coupled circuits with a coefficient of coupling on the order of one percent are often used.

Typical double-tuned transformers employ primary and secondary coils with a Q value of approximately 100. Critical coupling provides a secondary bandwidth of 25 kHz in this situation, and tighter coupling provides greater bandwidths. Greater bandwidth can also be obtained in the case of critical coupling if the primary and secondary windings are loaded with resistance. Although resistance loading reduces the output signal amplitude, it avoids excessive sag in the top of the response curve that results from tight coupling (overcoupling). Stagger tuning is still another approach that is sometimes used; this method of bandwidth increase entails tuning the primary and secondary to slightly different resonant frequencies. Note in Figure 4–14 that the maximum secondary current is obtained at critical coupling. Next, observe in Figure 4–15 that greater bandwidth at higher current output could be obtained by using coils with higher Q values. On the other hand, the resulting sag in the top of the response curve would be considered excessive. Again, if coils with low Q values were used, the output current would be decreased. An optimum Q value results in a comparatively high output current with a relatively flat-topped frequency response curve.

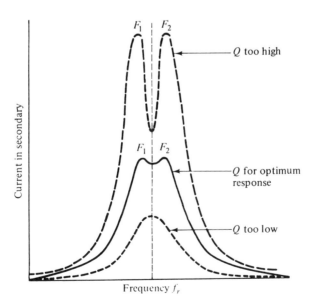

Figure 4–15 Frequency response curves for a double–tuned transformer with three different Q values.

Observe next the response-curve skirt contours in Figure 4–16. A single-tuned coupling circuit provides comparatively poor selectivity and little approximation to a flat peak response. On the other hand, a double-tuned coupling transformer with a product of coupling co-efficient and quality factor (kQ product) of 2 provides a reasonable approximation to flat-topped peak response, with a relatively rapid dropoff in response for frequencies past its maximum output point. Note that kQ products less than 2 result in progressively poorer selectivity and output-level characteristics. The response-curve contours depicted in Figure 4–16 are for single coupling circuits; improved frequency characteristics are obtained by cascading tuned stages. Thus, an advanced stereo FM tuner may use four or more cascaded, double-tuned coupling arrangements, with a corresponding improvement in selectivity.

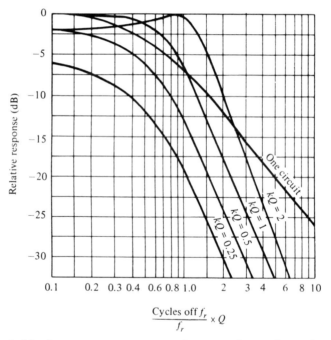

Figure 4–16 Response-curve contours for a single-tuned coupling circuit and for three different double-tuned coupling arrangements.

STEREO DECODER

Advanced stereo FM tuners use various types of multiplex decoders. A basic arrangement is shown in Figure 4–17; this switching-bridge decoder employs four diodes that operate as switches in response to

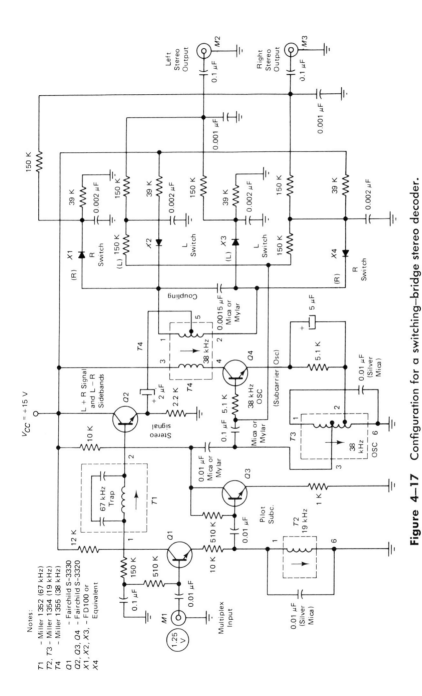

Figure 4–17 Configuration for a switching–bridge stereo decoder.

the peak-voltage output from the 38–kHz oscillator. This design develops the reconstituted 38–kHz subcarrier from a synchronized oscillator. In other words, the pilot subcarrier from the incoming signal is injected into the oscillator circuit. This is called a locked-oscillator design. A more sophisticated arrangement employs a 38–kHz oscillator with a phase-locked loop (PLL) configuration. This design references the 38–kHz oscillator frequency to the pilot subcarrier through a phase discriminator and low-pass filter arrangement. In turn, the 38–kHz oscillator is less likely to operate off frequency in the presence of high-amplitude noise bursts.

Another method of subcarrier regeneration is depicted in Figure 4–18. This design recovers the 19–kHz pilot subcarrier directly, and doubles its frequency through a full-wave rectifier circuit. In turn, the reconstituted 38–kHz subcarrier is developed for operation of the switching bridge. This method has the same limitation as the locked-oscillator arrangement in fringe-area reception. That is, high-level noise bursts tend to contaminate the 38–MHz output from the frequency doubler and thereby affect normal operation of the switching bridge. Therefore, most advanced stereo decoders employ a PLL design. Note that for maximum noise immunity, envelope detectors and matrixed configurations may also be operated with phase-locked loop circuitry. This PLL feature also helps to minimize distortion due to multipath reception under abnormal propagation conditions.

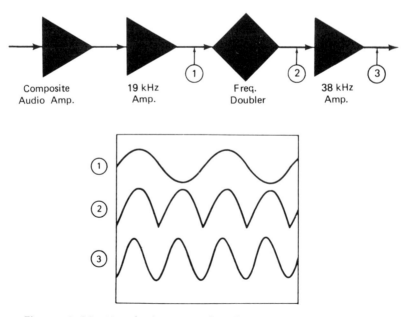

Figure 4–18   Key checkpoints in the subcarrier regenerator section.

INTEGRATED CIRCUITS

Integrated circuits are used extensively in advanced stereo FM tuners. Representative integrated-circuit (IC) packages are shown in Figure 4–19. A configuration for an IC FM IF strip appears in Figure 4–20. The first IF stage operates as a differential amplifier; its internal circuitry is depicted in Figure 4–21. Except for the temperature-stabilizing diodes in the base-emitter circuit of Q3, this is a familiar differential configuration with a constant-current source. However, the second IC

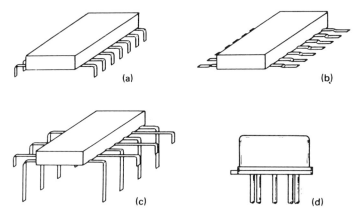

(a)   (b)

(c)   (d)

Figure 4–19 Typical flat-pack and TO-5 style IC packages. (**a**) In-line through-board mounting type; (**b**) in-line surface-mounting type; (**c**) quad-formed-lead mounting type; (**d**) transistor (TO-5) style package.

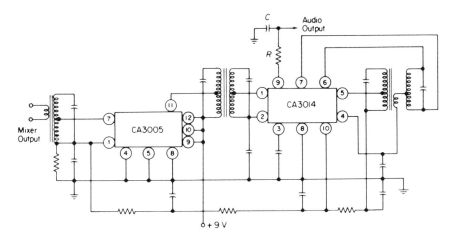

Figure 4–20 Configuration of an IC FM IF strip.

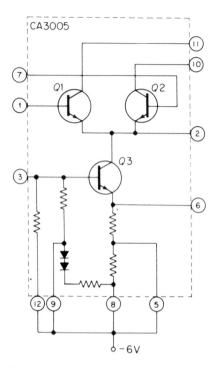

Figure 4–21   Internal circuitry of the first IC in Figure 4–20.

employs more complex internal circuitry, as shown in Figure 4–22. It consists of a multistage limiting amplifier, a voltage regulator, an FM detector, and an audio output stage. In turn, this IC can provide the nucleus of a complete IF strip. The first IC is included to provide additional gain. Thus, an IC, particularly the simpler forms, can be regarded as a high-gain transistor with a multiplicity of terminals. However, IC's with medium-scale integration (MSI) or large-scale integration (LSI) provide additional functions and cannot be regarded as merely high-gain transistors. For example, the IC depicted in Figure 4–22 has four functions and provides amplification at both IF and AF frequencies.

Note that the limiting amplifier section in the IC shown in Figure 4–22 comprises three cascaded differential amplifiers separated by emitter followers. Direct coupling is used throughout. Supply voltage for the first two differential amplifiers and their associated emitter followers is stabilized by transistor Q9 and the diode strings CR1 and CR2. Observe that the base voltage for Q9 is fixed by the constant voltage drop across this string of forward-biased diodes. Therefore,

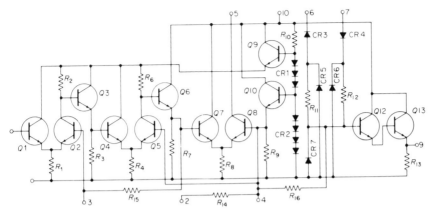

Figure 4–22 Internal circuitry of the second IC in Figure 4–20.

the emitter voltage of Q9 remains practically constant regardless of changes in supply voltage or current drawn by the input amplifier stages. Inasmuch as the base of Q7 is coupled to a regulated stage, the base voltage of Q8 must also be regulated in order to maintain stability of the differential pair. Observe the FM detector arrangement; this section requires only an external input transformer, all other demodulator components being contained in the IC. Detector diodes D3 and D4 work into an essentially resistive load. Audio output from the FM detector is obtained at the emitter of Q12; Q11 and Q12 operate as a Darlington pair.

In Figure 4–20, the IF signal from the mixer output circuit is applied to the base of Q2 (Figure 4–21). In turn, the IF signal is amplified and is applied to the second IC through a double-tuned IF transformer. The voltage gain from the input of the first IC to the input of the second IC is approximately 25 dB. Next, the second IC provides a gain of about 70 dB, so that the total gain of the IF strip is approximately 95 dB. The audio output from terminal 9 is applied to the deemphasis network RC. This is a low-pass filter that reduces the high-frequency response of the audio channel. Deemphasis is required for high-fidelity reproduction because the FM signal is preemphasized at the FM broadcast station. Preemphasis is provided to maximize the signal-to-noise ratio. In other words, the higher audio frequencies have comparatively low amplitude and may be masked by noise unless preemphasis is employed. Preemphasis and deemphasis characteristics are somewhat similar to treble-boost and treble-cut actions of tone controls.

AUDIOSCOPE MONITOR

Elaborate high-fidelity advanced stereo-quad systems may include a specialized oscilloscope-vectorscope section built into the integrated amplifier for monitoring channel signals. An audioscope is usually provided with controls to display left-front, right-front, left-rear, and right-rear signals with a linear time base. A typical channel waveform is shown in Figure 4–23. Triggered sweep action is provided so that the screen pattern is automatically synchronized. An audioscope can usually be switched for vector displays. A built-in audio generator may be provided with a typical frequency range of 20 Hz to 20 kHz. Most audioscope arrangements, however, require external stereo or quadraphonic test signals, for which demonstration tapes or records are a convenient source. Note that an audioscope usually has a 3–inch screen, whereas conventional oscilloscopes generally have 5–inch screens.

When switched on for quadraphonic vectorscope operation, an audioscope screen has four quadrants of display. In other words, an FL vector will be displayed in the upper left-hand quadrant; an FR vector, in the upper right-hand quadrant; an RL (BL) vector, in the lower left-hand quadrant; an RR (BR) vector, in the lower right-hand quadrant. An audioscope is useful not only to display comparative signal amplitudes, but also to disclose the presence of serious distortion, as well as to indicate reception difficulties due to multipath reception of FM signals. Alternatively, an external oscilloscope may be connected at the FM-detector output in an integrated receiver, as in Figure 4–24. An advanced stereo integrated receiver with a built-in audioscope is illustrated in Figure 4–25.

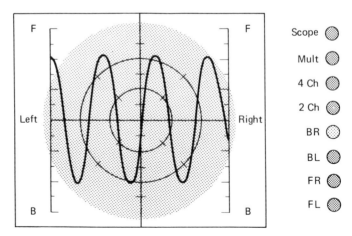

Figure 4–23 Typical display of an audio waveform on the screen of a built-in audioscope.

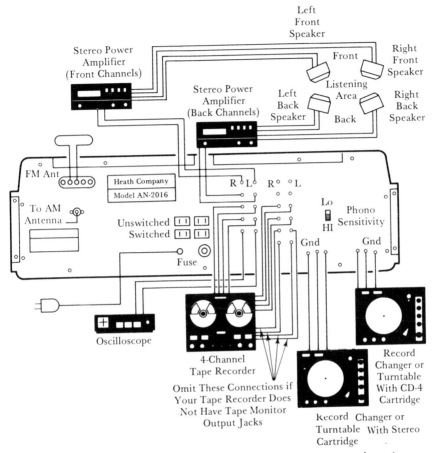

Figure 4–24 An oscilloscope can be connected to an integrated receiver to serve as an audioscope monitor.

Figure 4–25 An audioscope built into a high-quality integrated receiver. (Courtesy, Marantz)

5

High Technology
Tape Machines

Open-reel tape recorders and tape players have dominated the high-fidelity scene for over thirty years. Because open-reel machines are quite bulky and require threading of tape, cartridge machines are being widely adopted. Nevertheless, open-reel machines (Figure 5–1) are still preferred by the serious audiophile. An open-reel tape is ¼ inch wide and runs at constant speed past an erase head and a recording head, as depicted in Figure 5–2. The erase head, which is energized by

√ **Figure 5–1** An advanced open-reel stereo tape recorder.

a high-frequency sine-wave voltage, can remove any previous recording from the tape. The recording head is energized by the audio signal and impresses a varying magnetic field upon the coating of the tape. In turn, the coating becomes magnetized in accordance with the audio signal. A playback head is used to reproduce the recording, as shown in Figure 5–2(b).

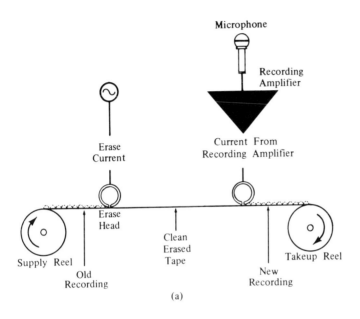

(a)

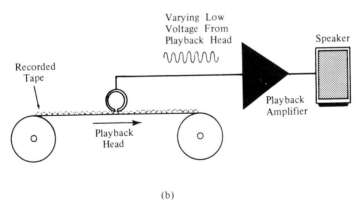

(b)

Figure 5–2 Functions of heads in a tape recorder. **(a)** Previously recorded audio signal may be removed by erase head; a new audio signal may be recorded by the recording head; **(b)** recorded audio signal is played back by the playback head.

126 High Technology Tape Machines

Open-reel tape machines provide tape-transport speeds of 7½, 3¾, and 1⅞ in/sec. A professional machine operates at 15 in/sec. Higher operating speeds permit reproduction of higher audio frequencies, with a trade-off in the length of tape required for a particular recording. Background noise (hiss) is minimized in open-reel machine operation. High-speed operation reduces the noise level. Open-reel magnetic tape employs four tracks, as pictured in Figure 5–3. Advanced stereo tape machines use the double Dolby system to suppress noise.

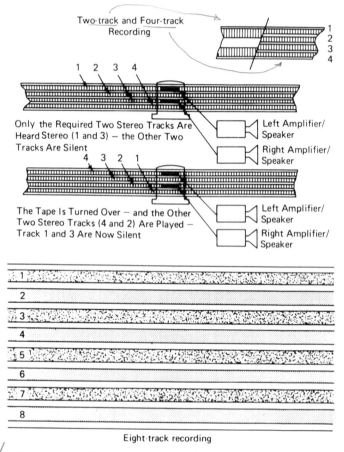

Figure 5–3 Two-, four-, and eight-track tape recordings.

In other words, Dolby bandwidth control is used during both the recording process and the playback process. Open-reel tape is supplied on reels as large as 10½ inches in diameter; 7½-inch reels are the most popular size. Sophisticated tape machines are designed with three heads, so that a recording can be monitored immediately after the magnetic pattern has been impressed on the tape, as shown in Figure 5–4. This facility avoids waste of time in the event that the recording process happens to malfunction. Simultaneous recording/monitoring is also termed A–B monitoring.

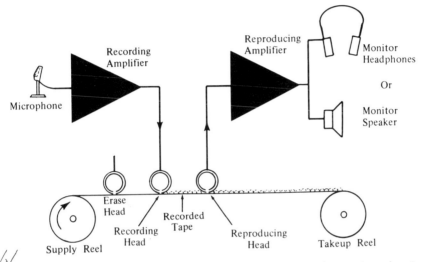

Figure 5–4   Three-head tape machine permits a tape to be monitored as it is being recorded.

TAPE MACHINE ORGANIZATION

A skeleton block diagram for a tape recorder is shown in Figure 5–5. It comprises a microphone, amplifier, tape transport, and bias oscillator. This bias oscillator generates a high-frequency sine-wave voltage in the range of 70 to 100 kHz. During the recording process, a comparatively high-level bias-oscillator voltage mixes with the audio signal to linearize the system transfer characteristic and thereby minimize distortion. Different kinds of recording tapes require different levels of bias voltage for optimum linearization. Advanced stereo tape machines have bias switches for the selection of correct bias levels for particular kinds of tape. A recording level meter is also provided so that the operator can see whether the sound level may be too high with resulting distortion owing to saturation of the magnetic tape. A recording meter will also indicate whether the sound level may be too low with resulting audible background noise.

Three principal types of magnetic tape are currently used: the conventional ferric trioxide tape, the chromium dioxide (CrO_2) tape,

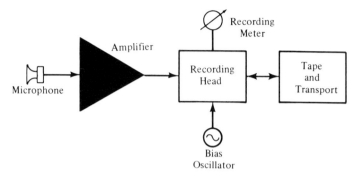

Figure 5-5 Skeleton block diagram for a tape recorder.

and the metal particle tape. Because each type requires a different level of bias voltage, advanced-stereo tape machines provide switching facilities for selection of bias voltage. Metal particle tape is the most recent development. Its performance surpasses that of the oxide-type tapes. This 'Metafine' (3M)* tape is composed of powdered metal (chiefly iron) instead of oxide compounds. It provides lower distortion, extended high-frequency response, and improved signal-to-noise ratio. Although playback of metal-particle tapes is feasible with conventional heads, modified recording and erase heads are required. A stronger erase current is also required owing to the higher coercivity and remanance parameters of metallic particle tape.

A-B MONITORING

A more complete block diagram for a tape recorder is shown in Figure 5-6. This arrangement employs separate recording and playback amplifiers—required for A–B monitoring. Note also that this assembly is more elaborate than a tape deck in that it provides both recording and playback facilities. Most tape decks are solely playback machines and lack amplifiers; however, some so-called tape decks also provide recording facilities. The arrangement in Figure 5–6 includes three heads. A few designs include a fourth head that applies a "crossfield" high-frequency bias independently of the record head. Note that some tape recorders are powered by a single motor, whereas others include two or three motors. When two motors are used, one drives the capstan

---

* 3M is a registered trademark of Minnesota Mining and Manufacturing Corp.

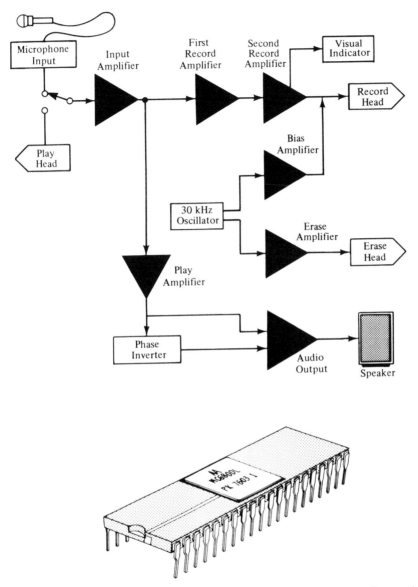

Figure 5–6 Block diagram for a tape recorder with separate recording and playback facilities.

and the other operates the take-up and rewind mechanisms. A three-motor machine uses one motor to drive the capstan; the other two motors rotate the feed reel and the take-up reel, respectively. Advanced stereo tape machines generally provide a vernier speed control for precise adjustment of transport speed; alternatively, a servo speed section may be employed.

RECORDING AND PLAYBACK CAPABILITIES

A typical open-reel machine of advanced design provides quadraphonic playback capability, but is limited to stereo recording capability. Some machines include automatic reverse playback of quarter-track stereo tapes; this feature eliminates the necessity for rethreading the reels for reverse playback. Four heads are employed in the automatic-reverse design. Note that three heads are essential for the aficionado who desires to record special effects such as echo or sound-with-sound. Although some tape machines record in only one direction, they will play back in both directions. Advanced-stereo designs occasionally include digital-logic units for automatic control of switching, braking, and clutch actions, thereby eliminating the hazard of breaking a tape because of mechanical stress in fast-forward or fast-rewind operations. Microprocessors available in a few types of tape machines allow highly precise control of motor speed; a quartz-crystal reference oscillator is generally included. Note that Dolby noise-reduction units are built into many open-reel machines and are optional as accessory units for other machines. Double-Dolby facilities are required for simultaneous A–B monitoring and noise reduction.

DIGITAL RECORDING AND PLAYBACK

Digital recording and playback are becoming more common in advanced stereo systems. Despite its relative costliness, this is the most effective method of reducing noise in tape-machine operation. In Figure 5–7, the signal-to-noise ratio of a recorded tape is equal to the peak signal voltage at maximum output divided by the peak noise voltage. Digital recording and playback techniques effectively reduce the peak noise voltage to an extremely small value, whereas the peak signal voltage is unaffected. The chief digital mode being used consists of pulse-width modulation of the audio signal (PWM), as depicted in Figure 5–8. This is a digital sample-and-hold processing of the analog audio waveform. In other words, the audio waveform is sampled progressively and its instantaneous voltage values at the sampling intervals are converted into pulses. The width of each pulse is proportional to the instantaneous voltage value at the corresponding sampling interval. Next, the PWM waveform is clipped, and thereby most of the noise fluctuations are removed, as shown in Figure 5–9. In turn, the signal-to-noise ratio of the clipped waveform impressed on the magnetic tape is maximized.

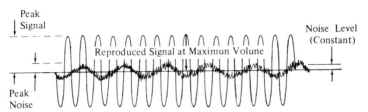

Figure 5–7 Visualization of signal-to-noise ratio.

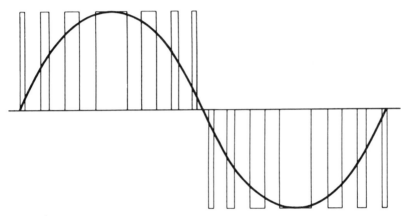

Figure 5–8 Pulse-width modulation of an audio signal.

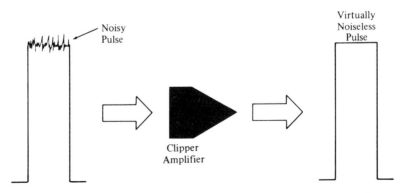

Figure 5–9 Clipper removes noise fluctuations that are superimposed on a digital pulse.

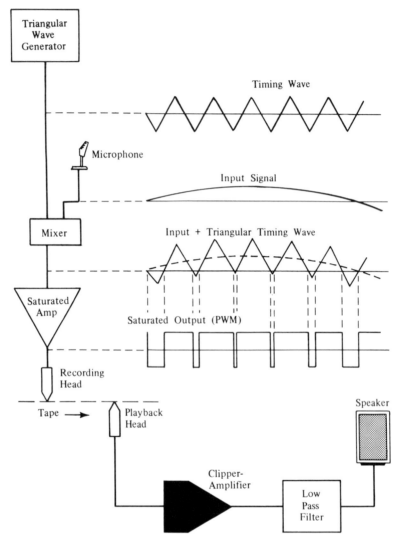

Figure 5-10   Generation of a PWM waveform for digital tape recorder.

Observe in Figure 5–10 that the output from the recording micro-phone is mixed with a triangular timing wave. This mixed waveform is then applied to a saturated amplifier (clipper-amplifier). Lower peaks of this timing waveform pulse the amplifier into operation, and the width of each output pulse is proportional to the instantaneous voltage of the audio waveform at the corresponding sampling interval. Thus, a high-fidelity recording is obtained with an unusually high signal-to-noise ratio. In the playback process, the reproduced pulses are first clipped to eliminate any tape hiss. The clipped, pulse-modulated waveform then passes through an integrating circuit, thereby reconstituting the original analog audio waveform for reproduction by the speaker system.

STEREO CASSETTE TAPE DECKS

Although cassette decks do not directly compete with open-reel tape machines in advanced stereo systems, the microgap heads for cassette machines have impressive performance capability. Cassette tape is approximately half the width (0.15 inch) of open-reel tape. Four tracks are provided in cassette tape recordings. A high-performance stereo cassette tape deck is illustrated in Figure 5–11. This machine provides a frequency response from 30 to 15,000 Hz when CrO_2 tape is employed. It includes a built-in Dolby noise-reduction section that provides a signal-to-noise ratio of 60 dB. This machine can also decode Dolby FM stereo broadcasts when connected into the tape-monitor circuitry that is included in most of the advanced stereo integrated receivers.

Figure 5–11 A high-performance stereo cassette tape deck. *(Courtesy, Radio Shack)*

Tape transport in a cassette deck tends to lack the high precision of advanced-stereo open-reel machines. However, some cassette decks are capable of impressive performance. Note that a small variation in tape speed modulates the frequency of reproduced tones. Comparatively rapid frequency modulation is termed flutter, and slow frequency modulation is called wow (see Chapter 1, Turntable Considerations). Either form of these mechanical distortions will impair reproduction fidelity.

Construction of a typical tape cassette is depicted in Figure 5–12. It consists primarily of a small pair of tape reels enclosed in a plastic housing. Cassettes are convenient because tape is automatically threaded between capstan and head when a cassette is inserted into the machine. Most cassette decks employ a single capstan for tape transport, but a few use two capstans, which improve precision of tape speed. A cassette provides about 40 minutes of playback time, whereas an open-reel machine with 7-inch reels provides several hours of playback time. A 10½-inch open reel has twice as much trackage as a 7-inch reel; a 5-inch open reel has half as much trackage as a 7-inch reel. Thus, even a 5-inch open reel provides considerably more playback time than does a cassette.

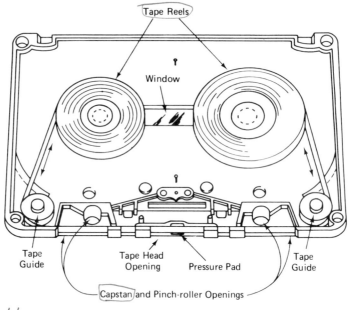

√√ **Figure 5–12** Construction of a typical tape cassette.

THREE-HEAD DESIGN

Three-head construction for purposes of A–B monitoring has become a trend in cassette decks. Some cassette decks use a combined record/play head; these are not included in the category of advanced stereo machines. All cassette machines transport the tape at a speed of 1⅞ in/sec. However, microgap head design permits extended high-frequency reponse. No tape machine can provide high-fidelity performance unless its heads are clean, correctly aligned, and periodically demagnetized. Worn heads should be replaced as required. Tape machines should be cleaned and lubricated at reasonable intervals, in accordance with the manufacturer's instructions.

AUTOMATIC LEVEL CONTROL

Cassette-deck and open-reel machines often include automatic level-control (ALC) facilities, as shown in Figure 5–13, which prevent tape saturation and resulting distortion of musical passages that have a wide dynamic range. Many designs provide a switch for defeat of the ALC function, if the operator so desires. Some machines employ amplified ALC, which provides a very large dynamic range without tape saturation by progressively compressing the higher amplitude peaks in the audio signal. However, wide-range automatic level control has certain disadvantages. When used inexpertly in problem environments, ALC can introduce objectionable distortion and noise. Another problem is that highly compressed musical passages tend to lose their "live" ambience. Also, an ALC system must be operated at a suitable preset level; otherwise, background noise will rise and fall noticeably as the ALC control voltage changes from a high value to a low value ("pumping" effect). Moreover, a variation of this "pumping" effect during certain kinds of musical passages may cause a periodic variation of background noise ("breathing" effect). In summary, ALC is not an unmitigated blessing to the neophyte stereophile, and this facility must be used judiciously.

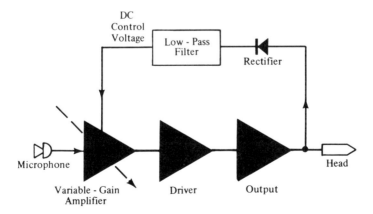

Figure 5–13   Automatic level-control arrangement.

VOLUME UNIT INDICATION

All advanced stereo tape machines have a pair of volume-unit (VU) meters for monitoring the recording levels of the L and R signals. This device prevents distortion due to tape saturation, or objectionable noise resulting from a poor signal-to-noise ratio. A typical VU meter is illustrated in Figure 5–14. In normal operation, the recording level is set so that the VU meters indicate zero for audio signals of average intensity. VU meters are calibrated in decibels-per-milliwatt (dBm) units. In other words, a dBm measurement is made with a dB meter that has a reference (0 dB) level of 1 mW in 600 ohms. A VU measurement is a dB measurement of a complex voltage waveform, such as a vocal or a musical passage. In turn, VU measurements are made with a special type of dB meter that has a reference level of 1 mW in 600 ohms with a damping factor such that pointer overshoot is not greater than 1.5 percent in response to a suddenly applied voltage. Thus, a dBm measurement applies to a steady sine-wave signal wherein damping is inconsequential; on the other hand, a VU measurement applies to a transient waveform such as a vocal or musical signal wherein damping must be appropriate in order to obtain a valid indication of the prevailing power level.

Figure 5–14 A VU meter. (Courtesy, Simpson Electric Co.)

Nearly all tape machines use dynamic microphones such as the one depicted in Figure 5–15. These are low-level transducers with output impedances in the range of 150 to 600 ohms. A typical dynamic microphone develops an output of approximately 300 microvolts (μV). As seen in the illustration, a wind screen of fine wire mesh is ordinarily mounted over the front of the housing. This screen reduces interference that can be caused by gusts of wind; it also attenuates "popping" sounds that can occur when a vocalist sings directly into the micro-

√ **Figure 5–15** Typical dynamic microphone.

phone. Comparatively costly microphones are often demanded by advanced stereophiles; however, many moderately priced microphones meet all but the most discriminating demands for high-fidelity reproduction. Microphones generally have three-terminal connectors, as outlined in Chapter 8.

In the past, "crystal" microphones were widely used with tape recorders. This type of microphone is also termed a piezoelectric or ceramic transducer. Despite its comparatively high output voltage, this design has been unable to realize high-fidelity response in feasible engineering versions. As a result, "crystal" microphones are not used in advanced stereo systems. The same general observations apply to conventional "sound-powered" microphones. Note that although a dynamic microphone is technically a sound-powered transducer, it is a product of high technology and has sophisticated features. A well-designed dynamic microphone meets high-fidelity standards that no conventional "sound-powered" microphone can reach. Note in passing that some so-called dynamic microphones also fail to meet high-fidelity standards. Moreover, a tape recorder that is designed to operate with a comparatively high-level microphone may fail to record satisfactorily with a low-level microphone, even though the latter provides high-fidelity output. In other words, there must be a reasonable "match" between the microphone output voltage and the rated input voltage for the recorder.

Condenser microphones, as illustrated in Figure 5–16, are capable

√
Figure 5–16 Condenser microphone provides highest fidelity.

of highest fidelity. This design has a comparatively low inherent level and also a very high internal impedance. Accordingly, it is always fabricated with a built-in high-fidelity preamplifier that provides an output level comparable to, or greater than that of a dynamic microphone, and also an output impedance that matches the input impedance of standard tape machines. In the opinion of many audiophiles, condenser microphones can optimize advanced stereo systems.

6

Specialized
Stereo Equipment

GENERAL CONSIDERATIONS

Advanced stereo systems often include various types of specialized equipment. As an illustration, a wireless remote control (Figure 6–1) can turn a tape machine on and off, change tracks, adjust volume, and operate the fast-rewind or fast-forward mechanisms. Young people often supplement a stereo system with a color organ, such as the one pictured in Figure 6–2. A typical color organ contains groups of lights with three different colors, energized through low-pass, bandpass, and high-pass filters. Filter outputs are applied to silicon controlled rectifiers (SCR's). Thus, when bass tones exceed a preset threshold level, the group of red lights may flash; when midrange tones exceed a threshold level, the group of green lights may flash; when treble tones exceed this level, the group of blue lights may flash. More elaborate color organs employ four groups of colored lights energized through four active filters.

When a stereo system is used by dancers at a party, young people often like to supplement the sound output by a strobe light such as the one shown in Figure 6–3. This type of light creates "stop-action" effects in the dance area by "freezing" motion at rapid successive intervals. The high-intensity xenon strobe light in this device flashes at an adjustable speed from three to ten times per second. A chrome reflector inside the cabinet provides high-intensity light output. A stereo system may also be supplemented with an audio mixer such as the one pictured in Figure 6–4 that can add "live" special effects to reproduced musical passages. The illustrated mixer handles four channel inputs, amplifies the inputs, and compresses unduly high-amplitude sound peaks. The compressor circuit provides fast attack, so that very rapid reduction in gain is available to prevent overload of the following amplifiers and speakers. This fast attack is followed by a delayed release time back to the initial gain level, so that a constant level of output is maintained during short pauses in musical or vocal passages.

Figure 6–1   Wireless remote control may be utilized during tape recording and editing operations.

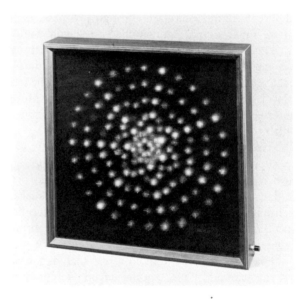

Figure 6–2   A color-organ display. *(Courtesy, Heath Co.)*

Figure 6–3   A wide-angle xenon strobe light. *(Courtesy, Radio Shack)*

Figure 6–4   A basic four-channel audio mixer.

BUCKET-BRIGADE AUDIO DELAY

Acoustic dimension synthesizers often find application in advanced stereo systems. Generally, they operate in the way illustrated in Figure 6–5. Observe that bucket-brigade audio delay units are used to delay application of signal voltages to the LR and RR speakers. Alternatively, LF and RF signals can be fed through the delay units to the ambience speakers in the rear of the listening area. In either case, the delay interval permits simulated expansion of the listening area from front

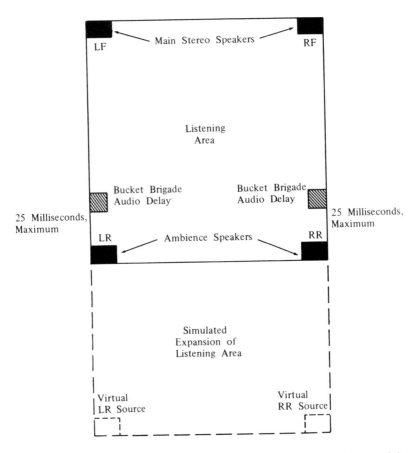

Figure 6–5 Simulated expansion of listening area by use of audio del units with LR and RR speakers.

to rear. As sound is reflected and re-reflected in a concert hall, it loses high-frequency energy progressively. As a result a reflected sound wavefront usually undergoes a treble cut. Moreover, sound originating from the rear of a listener tends to have treble attenuation compared with sound originating in front of the listener. This frequency discrimination is based on the physical characteristics of the human ear. Because the sound energy from the rear of a listening area is deficient in treble components, the audio delay units depicted in Figure 6–5 should provide proportionally more bass reproduction than treble reproduction.

Response of a typical time-delay ambience system to a tone burst is shown in Figure 6–6; the lower trace in the oscilloscope display indicates the time delay that is involved. Commercial delay units allow

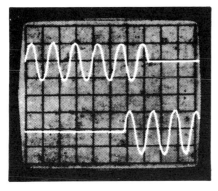

Figure 6–6   Tone-burst test of a time-delay ambience system, displayed on the screen of a dual-trace oscilloscope.

adjustment of the time-delay interval, so that various acoustic environments can be optimized. This system changes the audio signal into a chain of digital pulses by means of an analog-to-digital (A–to–D) converter. In other words, the instantaneous amplitudes of the audio waveform are sampled at rapidly successive intervals and are thereby converted into digital pulses with proportional amplitudes. This A–D conversion is accomplished by means of sample-and-hold circuitry. Then, the train of digital pulses is passed through a bucket brigade (shift register); the propagation time inherent in the shift register is equal to the delay time provided by the ambience unit. Sampling and shift-register timing are controlled by a digital clock with an adjustable rate of operation.

A typical time-delay ambience unit uses a 512-stage bucket-brigade device, pictured in Figure 6–7. A partial skeleton diagram for the device is shown in Figure 6–8. P-channel silicon gate technology is used with chains of tetrode-type MOS transistors in an integrated circuit. Note that this device can achieve tremolo, vibrato, and/or chorus effects in electronic musical instruments, in reverberator units, in complex waveform phase shifters, and in time compression and voice scrambling processors for communication systems. When comparatively long delays are desired, as in echo effects, two or more devices may

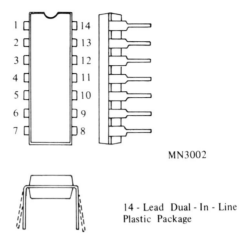

MN3002

14 - Lead Dual - In - Line
Plastic Package

Figure 6–7 A 512-stage bucket-brigade device for a time-delay ambience unit. *(Courtesy, Radio Shack)*

be cascaded, and when the total delay between front and rear sources exceeds 25 milliseconds, the listener will begin to experience an echo effect, instead of sound coloration.

The signal delay time versus clock frequency is graphed in Figure 6–9. Thus, a single device operating at a clock frequency of 10 Hz provides a time delay of approximately 25 milliseconds. Or, if the clock frequency is increased to 500 kHz, delay time becomes 0.5 milliseconds. As would be expected, two devices operating in cascade develop twice the delay time of a single device. Frequency response of the bucket-brigade device is a function of the clock rate, as depicted in Figure 6–10. Thus, a clock rate of 100 kHz accommodates an audio frequency of 10 kHz; a clock rate of 40 kHz (typical ambience rate) accommodates an audio frequency of 5 kHz; a clock rate of 10 kHz accommodates an audio frequency up to 1.5 kHz. This progressive attenuation of treble frequencies versus delay time is characteristic of normal sound coloration in typical acoustic environments, as noted previously. The device meets advanced stereo fidelity requirements in that it has a total harmonic distortion of less than 0.4 percent in typical applications.

A configuration for a time-delay ambience unit using the bucket-brigade device is shown in Figure 6–11. (pp. 148–49) The unit includes a reverberation output channel with adjustable rate and depth. L and R stereo inputs are accommodated; an insertion loss of 8.5 dB is imposed at a clock rate of 40 kHz and is compensated in the stereo system by advancing the preamplifier volume controls as required. The clock

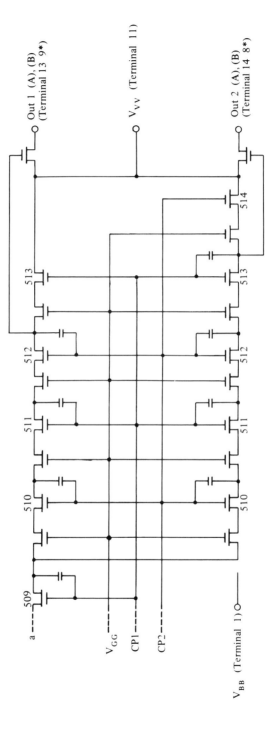

Figure 6–8 Partial skeleton diagram showing circuitry for transistors 509 through 513 in the IC time-delay ambience unit. (*Courtesy, Radio Shack*)

145

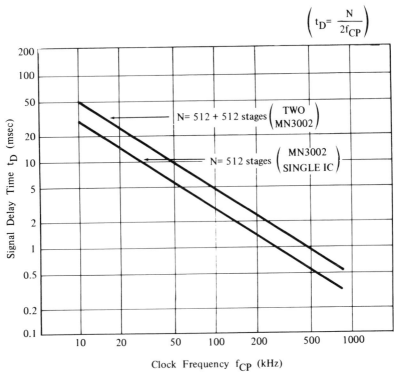

$$\left(t_D = \frac{N}{2f_{CP}}\right)$$

N= 512 + 512 stages $\left(\begin{array}{c} \text{TWO} \\ \text{MN3002} \end{array}\right)$

N= 512 stages $\left(\begin{array}{c} \text{MN3002} \\ \text{SINGLE IC} \end{array}\right)$

Clock Frequency f_{CP} (kHz)

Figure 6–9 Signal delay versus clock frequency for one device and for two devices operating in cascade. *(Courtesy, Radio Shack)*

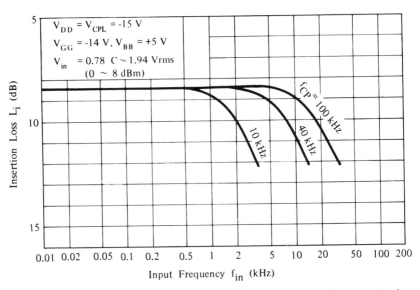

$V_{DD} = V_{CPL} = -15$ V

$V_{GG} = -14$ V, $V_{BB} = +5$ V

$V_{in} = 0.78$ C ~ 1.94 Vrms
(0 ~ 8 dBm)

$f_{CP} = 100$ kHz

10 kHz

40 kHz

Input Frequency f_{in} (kHz)

Figure 6–10 Audio frequency response versus clock rate. *(Courtesy, Radio Shack)*

146

section in the lower left-hand corner of the diagram is basically a free-running multivibrator having a type 555 IC and a bipolar transistor. The bucket-brigade device operates in combination with input and output buffer amplifiers contained in an associated IC (IC1). A regulated power supply provides outputs of 20.6, 15, and 1 V to the three devices used in the bucket-brigade audio delay unit.

STEREO POWER OUTPUT METER

A few stereo power amplifiers have output meters that can monitor power outputs. An accessory power-output meter unit is illustrated in Figure 6–12. Each scale indicates up to 100 watts, and the entire unit is designed to operate in a 4–ohm or 8–ohm speaker system. The instrument circuitry is designed for peak-power indication. Response time is rapid, so that peak power values are indicated even on sharp pulses. This response characteristic is similar to that of modern volume-unit (VU) meters available with high-performance tape recorders. Power output meters help to prevent amplifier or speaker damage due to overload.

ACOUSTIC SCREENS

An acoustic screen is a sheet of sound-absorbent or sound-reflective material that is used to modify the acoustics of a listening area. A screen may be plane or curved, or folding as in Figure 6–13. Bass screens are large compared to treble screens. Acoustic screens are usually painted with decorative designs to match room decor. One or more absorptive screens in a rumpus room or game room can reduce environmental reflectivity and reverberation time. On the other hand, one or more sound-reflective screens may be useful in a listening area with wall-to-wall shag carpeting, upholstered furniture, and large-area draperies. Acoustic screens can also "tailor" the treble profile of a listening area. Of the common shapes available, a round screen is most effective for any given area; a square screen is somewhat less effective, and an elliptical screen is least effective.

To develop an acoustic "shadow" on the farther side of an acoustic screen, the shortest dimension of the screen must be on the order of one wavelength relative to the screen's cutoff frequency (wavelength). As an illustration, a circular screen 6 feet in diameter will develop an acoustic "shadow" for frequencies down to approximately 200 Hz. A circular screen 10 feet in diameter will "shadow" sound wavefronts at frequencies as low as approximately 110 Hz.

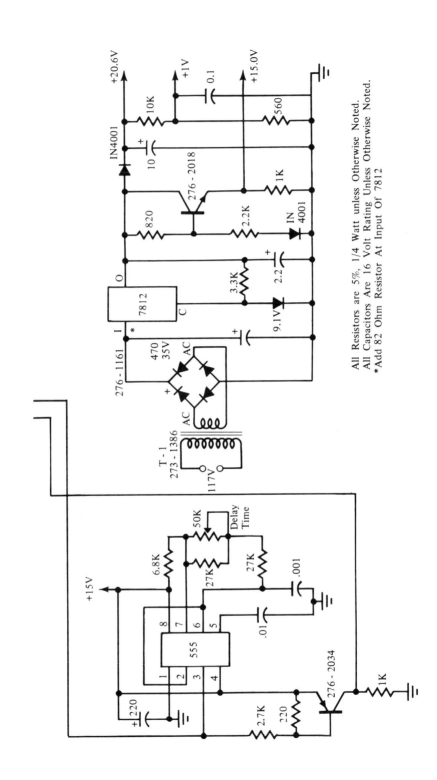

All Resistors are 5%, 1/4 Watt unless Otherwise Noted.
All Capacitors Are 16 Volt Rating Unless Otherwise Noted.
*Add 82 Ohm Resistor At Input Of 7812

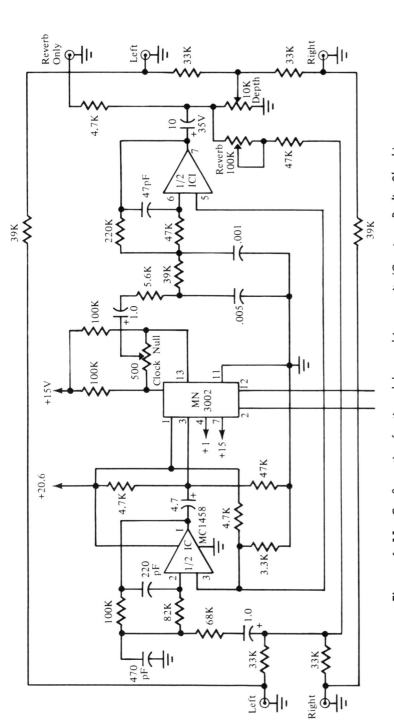

Figure 6–11 Configuration for time-delay ambience unit. (Courtesy, Radio Shack)

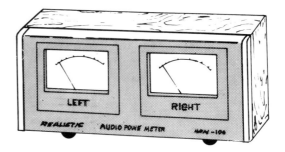

Figure 6–12   Stereo power-output meter. *(Courtesy, Radio Shack)*

Surface may be
highly absorptive

Surface may be
highly reflective

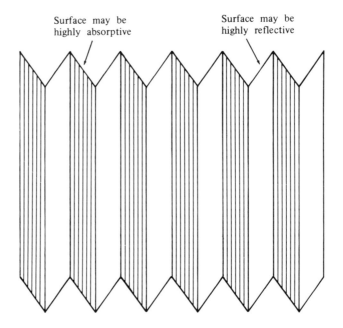

Figure 6–13   A folding type of acoustic screen.

This "shadow" action is based on acoustic diffraction and is independent of the absorption coefficient of the screen's surface. A treble "cut" occurs within the shadowed area; in other words, high-frequency sound waves are diffracted around objects to a lesser extent than are low-frequency sound waves. In the analysis of acoustic environments and in experimental procedures, it is helpful to work with both "dead" and "live" screens or with a "dead-live" screen that has been treated on one side to provide high reflectivity, whereas the other side is highly absorptive. Thus, one side of a "live-dead" screen may be a sheet of hard and polished plastic, and the reverse side fiberglass.

PROGRAMMABLE TURNTABLES

Some advanced stereophiles desire to operate a turntable with a remote transmitter and receiver so that it can be "programmed" from a hand-held push-button unit. A totally programmable turntable will play individual cuts in any chosen order; it will repeat-play or skip individual cuts; and it will change speeds for 33⅓ or 45 rpm discs. A typical wireless remote control transmitter operates on a 9–V battery and has a range up to 30 feet. The remote receiver plugs into the turntable housing. An array of controls and push buttons also present on the turntable housing permit manual operation.

BIPHONIC SOUND REPRODUCTION

Advanced stereo systems may employ biphonic sound reproduction. Outwardly this stereo arrangement uses either headphones or a pair of speakers, but technically, it is a sophisticated quadraphonic system. Although biphonic tape recordings can be played back directly into headphones, suitable processors must be used with stereo equipment for reproduction of biphonic sound by speakers. Because the region of good biphonic perspective is very limited for speaker reproduction, however, the listener must place himself at a certain point with respect to the speakers; otherwise, the listener will experience only a conventional stereo sound field. Biphonic sound reproduction provides positional clues in addition to the general directional clues that are provided by conventional stereo reproduction. For example, if a sound originates at "10 o'clock" with respect to the microphone in a recording studio, the general direction of the sound can be detected on stereo reproduction but the relative distance of the source cannot be detected.

Stated otherwise, the stereo listener cannot determine how far away a particular sound originated, but he can determine the direction of the sound with reasonable accuracy (provided that the sound originated from the front of the recording studio). Thus, a stereo system introduces directionality that is lacking in a monophonic system, but it takes considerable sophistication to produce complete directional clues and positional clues. A quadraphonic system provides complete directional clues; so that sounds originating at the rear of the recording studio are not confused with sounds originating at the front of the recording studio. Primarily, these clues consist of proportionally greater or lesser outputs from the rear speakers than from the front speakers, plus comparatively greater output from one of the rear (or from one of the front) speakers than from the other.

Next, in order to determine the relative distance of a sound source in a particular direction, the listener must be able to perceive subtle acoustic factors like the ones that provide positional clues for sound sources in "real life." This is the province of biphonic sound. One important acoustic factor the ear senses is the relative phase of a sound wavefront when a sound originates to one side of the listener. Another factor that provides positional information is the diffraction that a sound wavefront undergoes owing to the "shadow" effect of the listener's head in the sound field. This diffractive action becomes significant for tones with frequencies above 700 Hz. Basic mono, stereo, quad, and biphonic acoustic factors are pictured in Figure 6–14. Observe in Figure 6–14(a) that in mono reproduction all sounds appear to originate at the speaker. Only relative loudness information is provided directly, although now and then coloration of sound in the recording studio may provide indirect directional clues. Thus, the listener occasionally may be able to make an "educated guess" concerning the position of a sound source, even in mono reproduction.

Both loudness and forward-directional information are provided in stereo sound reproduction, as depicted in Figure 6–14(b). Both loudness and omnidirectional information are conveyed in quadraphonic sound, as shown in (c). If positional information is to be included, biphonic sound reproduction must be employed, as pictured in (d). Note that when earphones are used in biphonic sound reproduction, the acoustic environment resembles an anechoic chamber. This is an optimum acoustic situation for the reproduction of positional information, inasmuch as the perceived sound is thereby uncolored by listening-area acoustics. The acoustic factors that are provided in biphonic sound signals add dimension (positional clues) to the directional information of quadraphonic sound signals. Thus, biphonic sound reproduction is also termed dimensioned stereo reproduction. Note that biphonic recordings require sophisticated biphonic microphones.

Commercial biphonic tape recordings can be played back directly into headphones. On the other hand, when biphonic tape recordings are to be reproduced by a pair of speakers, the signals from the tape player must be suitably processed before they are applied to the speakers in order to achieve complete separation of the outputs from the L and R speakers. Therefore, acoustic crosstalk must be cancelled out by means of out-of-phase signal components at the listening position. Biphonic signal processors provide acoustic crosstalk cancella-

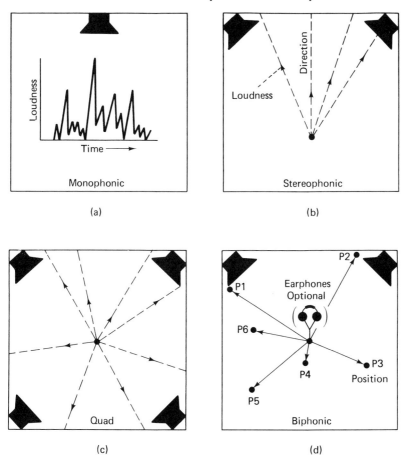

Figure 6–14   Basic acoustic factors. (**a**) Monophonic system provides loudness information; (**b**) stereophonic system provides loudness and forward-directional information; (**c**) quadraphonic system provides loudness and omnidirectional information; (**d**) biphonic system provides loudness, omnidirectional, and positional information.

tion components that are introduced into the biphonic L and R signals before they are applied to the speakers. Another requirement for speaker reproduction of biphonic sound is that the listener position himself at a spot precisely in front of the speakers where crosstalk is exactly cancelled. Maximum biphonic perspective occurs at this spot, and the resulting perception of sound is more realistic than in quadraphonic sound reproduction. In other words, the listener can estimate the distance of a sound source from the microphone at the recording studio with almost as much accuracy as if he were occupying the position of the microphone. Comparative directions and locations of a sound source processed by mono, stereo, quad, and biphonic systems are summarized in Figure 6–15.

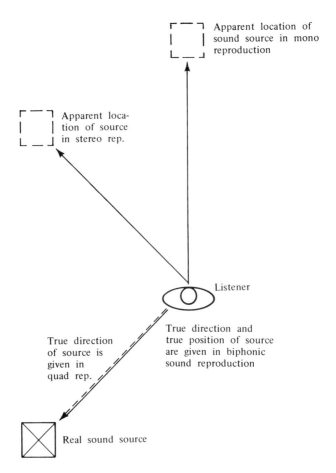

Figure 6–15 Comparative directions and locations of a sound source processed by mono, stereo, quad, and biphonic systems.

7

Advanced Stereo Tests and Measurements

Tests are distinguished from measurements in that the former determine qualitative characteristics of stereo components, whereas the latter determine quantitative values. Electrical, electronic, and acoustic characteristics and values are checked in advanced stereo operations. For example, an electrical test of a speaker may be made by connecting a flashlight cell across its terminals in order to determine whether it will respond with a "click", or whether it might be "dead." An electronic check of turntable speed may be made with a strobe light. Or, the treble profile of a listening area may be plotted by walking across the floor with a sound-level meter while the stereo speakers are energized with a constant frequency, such as 8 kHz. Connecting a flashlight cell across the terminals of a speaker is an example of a go/no-go type of test. A strobe check of turntable speed is an attempt to verify a specified operating value. The procedure of determining the treble profile of a listening area represents a quantitative measurement.

Consider the re-recording merit test for a pair of tape machines, pictured in Figure 7–1. In this example, a tape deck applies an input signal to a tape recorder with A–B test facilities; the re-recorded material is monitored with the speaker. After the first re-recording, the reels are interchanged between the deck and the recorder, and a second re-recording is made from the first re-recording. This procedure is repeated a number of times, until the reproduced sound from the monitor speaker deteriorates to the point that it is no longer acceptable. The merit value of the deck-recorder system is represented by the number of re-recordings that were made. This is an example of a

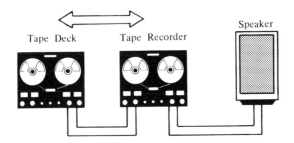

Figure 7–1   Re-recording merit test of tape equipment.

semi-quantitative test. In other words, if a certain deck-recorder system is capable of n re-recordings, and another deck-recorder system is capable of 2n re-recordings, it is concluded that the latter is "twice as good" within the framework of the test method.

In general, speakers and microphones have appreciably more frequency distortion and harmonic distortion than do tape machines. A semi-quantitative test of speaker-microphone system merit can be made as shown in Figure 7–2. This is a re-recording test procedure in which the speaker is frequently the "weakest link in the chain";

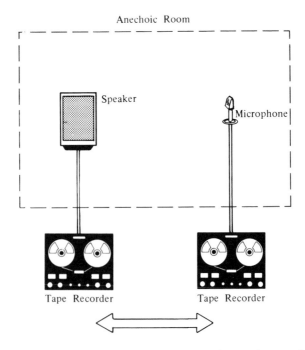

Figure 7–2   Re-recording merit test of speaker and microphone.

it is primarily a comparative test of speaker merit. The microphone is energized by the output from a speaker in an anechoic room. After the first re-recording, the reels are interchanged between the two tape recorders, and a second re-recording is made from the first re-recording. This procedure is repeated several times, until the reproduced sound from the speaker is no longer acceptable. The merit value of the system is denoted by the number of re-recordings that were made. Since the speaker ordinarily contributes most of the distortion in the system, this is basically a comparative merit test for different speakers.

SOUND-LEVEL METER

A sound-level meter such as the one illustrated in Figure 7–3 is the main instrument required to check an acoustic environment. It consists of a high-fidelity condenser microphone, a battery-powered integrated-circuit amplifier, and a meter calibrated in decibels. Six sound-level ranges are provided, from 60 dB to 126 dB, that are referenced to the standard acoustic zero dB level of 0.002 microbar (μ bar). Both A and C weightings are provided for the indicating scale. The A weighting is almost uniform over the frequency range of 32 Hz to 8 kHz and indicates the physical level of sound energy in the listening area. On the other hand, the C weighting provides a bass-attenuation characteristic that is similar to the hearing characteristic of the human ear. In other words, the scale indication is approximately proportional to

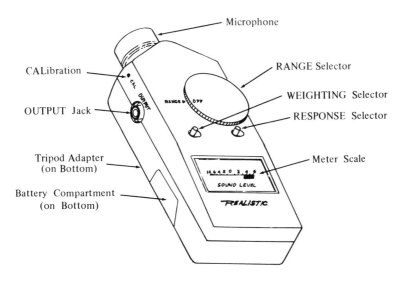

Figure 7–3 A high-performance sound-level meter. *(Courtesy, Radio Shack)*

phon units, when the A–weighting mode of operation is used (see Figure 7–4). The sound-level meter provides a choice of fast or slow response; fast response indicates peak values of transient acoustic waveforms, whereas slow response indicates average values of acoustic waveforms.

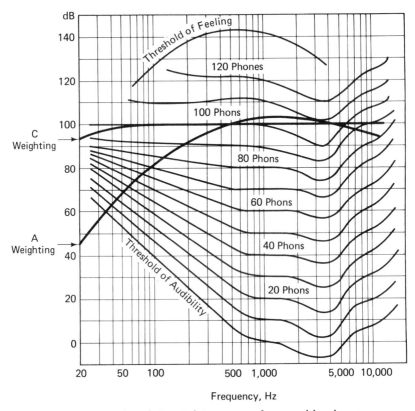

Figure 7–4   A and C weighting curves for sound-level meter.

It is often advantageous to mount a sound-level meter on a tripod, so that hand noise and sound reflections from the operator's body can be eliminated. Tripod mounting is also convenient when a sound-level meter is used with auxiliary equipment such as a tape machine or an oscilloscope. Note that when sound waves are coming chiefly from one direction, the scale indication on a sound-level meter may be significantly influenced by reflections from the operator's body. If the sound-level meter is held directly between the operator and a sound source, an indication error of several dB can be anticipated in the frequency range above 100 Hz. It is helpful to position a sound-level meter so that the principal sound wavefront arrives at right angles to the front of the microphone. However, the microphone used in the illustration is primarily omnidirectional, so that sound levels can be measured reliably from sources in any direction, provided that the wavefront is not distorted by reflections from the operator's body.

Note that the size, shape, and furnishings of a listening area can have a profound effect on the performance of an advanced stereo system. A "hard" room with bare surfaces tends to exaggerate bass response and gives treble tones a "honking" timbre. On the other hand, a "soft" room amply furnished with curtains and carpets tends to inhibit bass response and adds a shrill timbre to high-frequency tones. Depending on speaker locations, acoustic standing waves may develop in a listening area, and create a peaky and eccentric acoustic profile. Sound distribution in a listening area is determined as shown in Figure 7–5. An audio generator energizes a speaker at a chosen location in the room, and, the sound-level meter is moved across the room along a number of paths spaced progressively closer to the speaker. This test is usually made at a low bass frequency, a midrange frequency, and a treble frequency. The audio signal generator should provide a good sine waveform for optimum testing of sound distribution. A high-performance generator is illustrated in Figure 7–6.

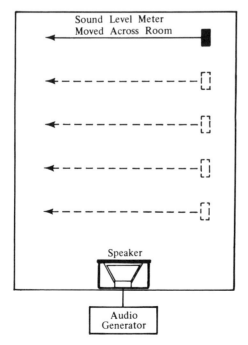

Figure 7–5   Determination of sound distribution in a listening area.

Figure 7–6   A good quality audio signal generator with an output level meter and a wide-range attenuator. *(Courtesy, Leader Electronics)*

Most listening areas have pronounced resonant frequencies that are easily identified by means of the test method shown in Figure 7–7. An audio generator is connected to a speaker in a chosen location, and a sound-level meter is placed at the listening position in the room. The audio generator is then tuned slowly through a frequency range of 20 Hz to 20 kHz, as the sound-level meter response is monitored. As an acoustic resonant frequency is approached, the meter reading will start to rise rapidly; it attains a maximum value at the peak resonant frequency, and then falls through progressively lower values at frequencies on the other side of resonance. Several prominent resonant frequencies are likely to be found in a typical stereo listening area. There are two basic approaches to acoustic resonant-frequency problems. Many stereophiles use a frequency equalizer in the audio channel to compensate for acoustic resonances. Alternatively, acoustic treatment may be accorded the listening area by means of draperies, floor rugs, and acoustic screens, all of which affect acoustic resonant frequencies. In some situations, a combination of acoustic treatment and frequency equalization will provide optimum sound reproduction.

MEASUREMENT OF ACOUSTIC SEPARATION

Although related, electronic separation, acoustic separation, and psychophysical separation are quite different processes. Unless headphones are used, acoustic separation is always less than the electronic separation in a stereo system; *headphones provide 100 percent acoustic separation.*

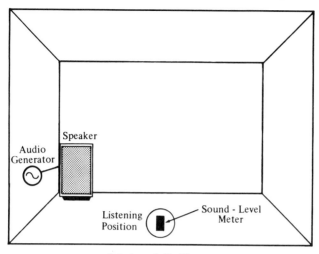

Interior of Residence

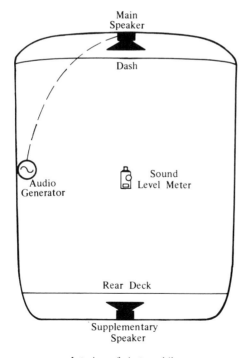

Interior of Automobile

Figure 7–7   Measurement of resonant frequencies in a listening area.

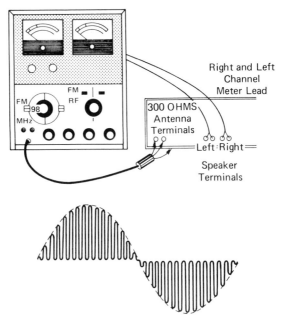

Figure 7–8 Electronic separation is measured with a stereo-multiplex generator. *(Courtesy, Sencore)*

Figure 7–8 shows the electronic separation provided by the multiplex decoder of an FM tuner being measured with a stereo-multiplex generator and an indicator such as a pair of dB meters (often built into the generator). A separation of 30 dB between the L and R channels in the decoder meets the requirements of an advanced stereo system. Next, consider the measurement of acoustic separation in a listening area, as depicted in Figure 7–9. Here, acoustic separation equals the differences in readings on two sound-level meters located at the listening position, and approximately six inches apart to simulate the spacing between the listener's ears. The L and R speakers are alternately energized by a 1–kHz tone from an audio generator, and the corresponding readings for the two sound-level meters are observed. A small acoustic separation may occur at higher test frequencies, but zero acoustic separation occurs at 1 kHz. In many acoustic environments, almost zero acoustic separation occurs at the listening position (i.e., both sound-level meters indicate the same value, whether the sound source is from the L speaker or from the R speaker).

Note carefully that although the foregoing test may indicate zero acoustic separation, pronounced psychophysical separation is experi-

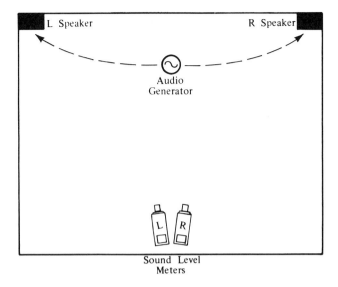

Figure 7–9   Measurement of acoustic separation in a listening area.

enced in most acoustic environments. Psychophysics is the study of interrelations between physical processes and mental processes. Psychophysical separation of L and R sound sources in a listening area is a function of the listener's binaural perception of sound. In other words, a sound from the L speaker arrives at the listener's left ear a small fraction of a second before it arrives at his right ear. As well, there is a slight phase difference between the sounds perceived by the listener's left and right ears. As a result, a sound wave from the L speaker will be perceived directly by his left ear, whereas this sound wave will be slightly "shadowed" by the listener's head on its path to his right ear. This "shadowing" effect attenuates the sound intensity to some extent, and it also imposes a treble cut. Attenuation and treble cutting become more pronounced in the higher audio-frequency range. The foregoing factors become directional clues for the listener, who experiences appreciable psychophysical separation in the listening area even though acoustic separation may be zero. These factors are summarized in Figure 7–10.

AUXILIARY INSTRUMENTS

In various types of tests, a sound-level meter may be supplemented to good advantage with other electronic instruments, as exemplified in

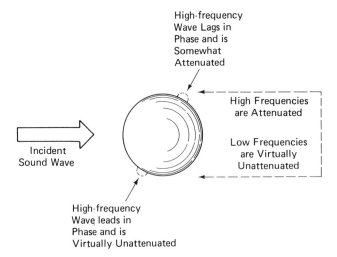

Figure 7–10   Diffraction of a sound wave by a sphere.

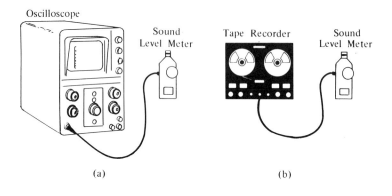

(a)   (b)

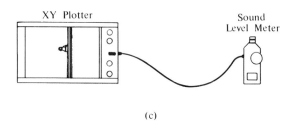

(c)

Figure 7–11   A sound-level meter may be supplemented by other electronic instruments. **(a)** Oscilloscope indicator; **(b)** tape recorder; **(c)** XY plotter.

Figure 7–11. For example, an oscilloscope connected at the output of the amplifier in a sound-level meter displays the waveform of the sound energy reproduced by the microphone. The sequence of sounds incident upon the microphone during tests with a sound-level meter can be recorded on tape and later the tape recording may be analyzed by display of the recorder output on an oscilloscope screen. When a permanent record of the sound waveforms is desired, an XY plotter instead of an oscilloscope may be connected at the output of the amplifier in a sound-level meter. In the event that harmonic analysis is desired, the output from a sound-level meter may be connected to a spectrum analyzer, as shown in Figure 7–12. In this case, the frequencies and amplitudes of any harmonics (or other distortion products) will be displayed on the screen of the spectrum analyzer.

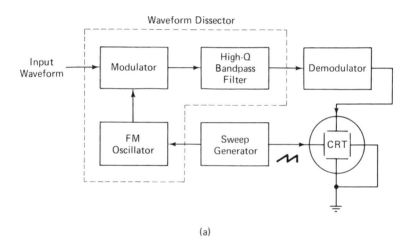

(a)

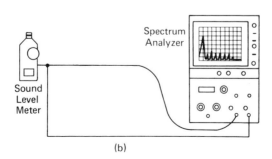

(b)

Figure 7–12   Harmonic distortion test with a spectrum analyzer. **(a)** Functional plan; **(b)** test setup;

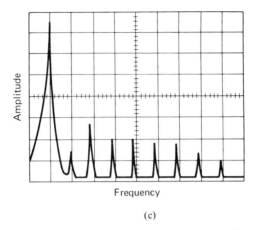

Frequency

(c)

Figure 7–12 *Continued:* **(c)** typical display.

VISUAL ALIGNMENT

Visual alignment (sweep alignment) of FM tuners is accomplished with the test setup depicted in Figure 7–13. Tuned circuits in the RF, IF, and demodulator sections are adjusted in accordance with procedures outlined in the appropriate service manual. After resonant circuits are properly adjusted, the specified frequency-response curves will appear on the oscilloscope screen (assuming that no malfunction in the receiver circuits requires correction). After an electronic technician gains considerable experience with visual-alignment equipment, he may proceed to check receiver alignment without having to consult the service manual. This generalized approach will sometimes lead to puzzling test results. As an illustration, suppose that the oscilloscope is connected into a circuit point following the limiter where there happens to be appreciable capacitive reactance. The trace and retrace excursions of the CRT beam will fail to lay over precisely, as anticipated, and double images will appear, as exemplified in Figure 7–14. The remedy, of course, is to select an oscilloscope take-off point at which the circuitry is resistive.

HARMONIC DISTORTION MEASUREMENT

Percentage harmonic distortion is measured with a harmonic-distortion meter, such as the one illustrated in Figure 7–15. The test setup is

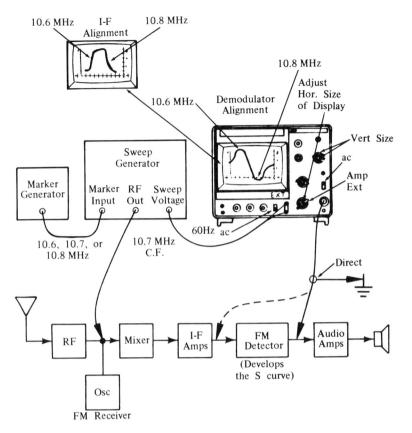

Figure 7–13 FM receiver sweep-alignment setup. *(Courtesy, B&K Precision, Div. of Dynascan Corp.)*

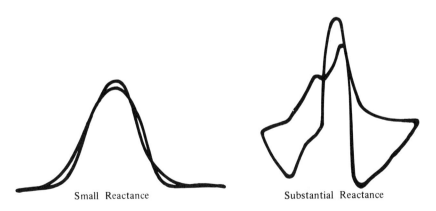

Small Reactance   Substantial Reactance

Figure 7–14 Trace and retrace fail to lay over precisely when there is appreciable reactance in the take-off (load) circuit.

167

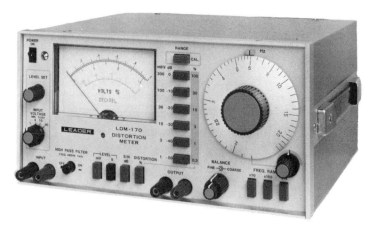

Figure 7–15 Appearance of a high-quality harmonic-distortion meter. *(Courtesy, Leader Electronics)*

pictured in Figure 7–16. It is essential to connect a power resistor of ample wattage rating and ohmic value to the amplifier output terminals in order to provide a normal load. Harmonic-distortion tests are usually made at 1 kHz. More informative test data will be obtained if distortion is also measured at 100 Hz and at 10 kHz. The amplifier is driven to

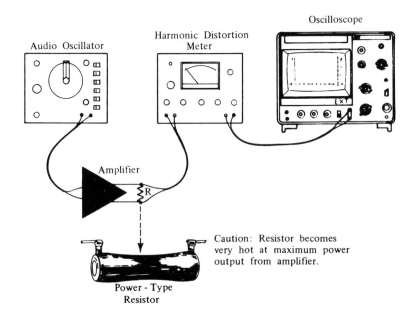

Figure 7–16 Harmonic distortion test setup.

maximum rated power output for a basic distortion measurement. However, in measuring the power bandwidth of an amplifier, the output from the audio oscillator is reduced to obtain half of maximum rated power output (70.7 units of maximum rated voltage output). An oscilloscope is informative in this test procedure because the displayed waveform will show whether the distortion products are primarily second, third, or higher harmonics. Moreover, a harmonic-distortion meter indicates hum and noise voltages as if they were harmonic voltages. In turn, an oscilloscope will show whether part of the meter reading should be attributed to hum and/or noise voltages.

Basic types of waveform distortion depicted in Figure 7–17 include: amplitude distortion (associated with harmonic distortion), frequency distortion (not associated with harmonic distortion), phase distortion (not associated with harmonic distortion), transient distortion (not characterized in terms of harmonic distortion), and time-reference or Doppler distortion (associated with harmonic distortion). Doppler distortion arises in acoustic situations—not in electronic circuitry. For example, if a large woofer is required to radiate treble tones, the high-frequency portion of the sound wavefront will be subject to Doppler distortion by the low-frequency portion of the sound waveform. For this reason, advanced stereo speaker systems are designed as three-way assemblies, in which a small tweeter speaker radiates the higher portion of the treble range. Note that a woofer with a whizzer cone is prone to considerable Doppler distortion.

SQUARE WAVE TESTS

Transient distortion can be checked with square waves, as shown in Figure 7–18. The amplifier under test is terminated in its rated value of load resistance, and an oscilloscope is used to display the reproduced square waveforms. Typical square-wave distortion patterns are depicted in Figure 7–19. The integrated (B&D combined) pattern often occurs at repetition rates in excess of the amplifier's transient capability. Conversely, the differentiated (C&E combined) pattern often occurs at repetition rates below the amplifier's transient capability. Most advanced stereo amplifiers are rated for almost undistorted 2–kHz square-wave reproduction; some are rated for square-wave reproduction at higher and at lower repetition rates. As noted in Figure 7–18, the rms power value of a square wave is equal to E^2_p/R_L, where E_p is the peak voltage of the square wave. Therefore, this relation must be used in calculating the power output from the amplifier when it is driven by a square-wave input voltage. Failure to observe this relation could result in damage to the amplifier under test, owing to overload.

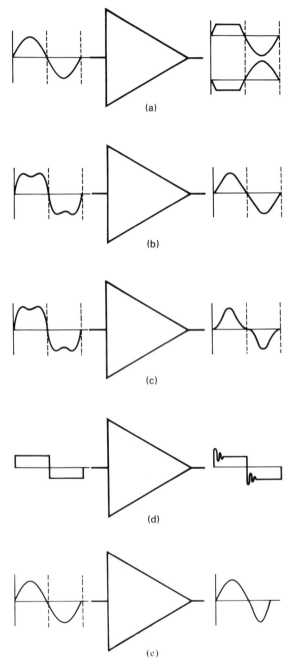

Figure 7–17 Basic types of waveform distortion. **(a)** Amplitude distortion; **(b)** frequency distortion; **(c)** phase distortion; **(d)** transient distortion; **(e)** time-reference (Doppler) distortion.

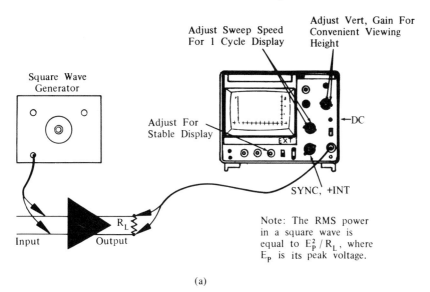

(a)

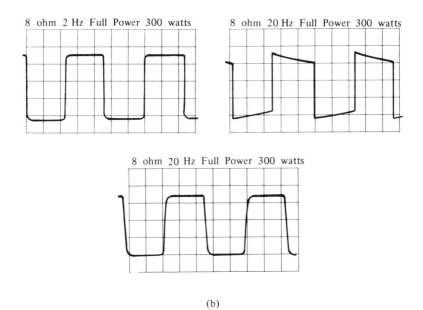

(b)

Figure 7–18   Transient test. **(a)** Test setup for square-wave test of amplifier response *(Courtesy, B&K Precision, Div. of Dynascan Corp.)*; **(b)** square-wave responses of a 300-watt power amplifier. *(Courtesy, S.A.E.)*

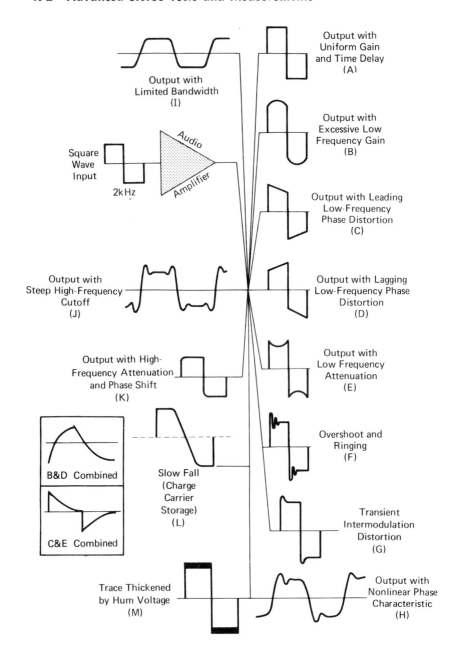

Figure 7–19 Basic audio-amplifier square-wave responses.

INTERMODULATION DISTORTION

Intermodulation distortion is related to harmonic distortion, in that both are caused by amplitude nonlinearity. However, harmonic distortion is measured with a single test frequency, whereas intermodulation distortion is measured with a pair of test frequencies (two-tone test signal). Intermodulation distortion involves the generation of new frequencies in the output from the amplifier; these spurious frequencies are equal to the sums and differences of integral multiples of the component frequencies present in the test signal. An intermodulation (IM) test signal is typically formed from a 60–Hz and a 6–kHz signal (both sine waves). When this two-tone signal is processed by an amplifier, any amplitude nonlinearity that is present will cause the 6–kHz signal to be amplitude-modulated by the 60–Hz signal. An IM analyzer contains filter and demodulator sections for picking out the amplitude-modulation components. These components are applied to a meter and are indicated on a percentage IM scale.

An intermodulation distortion test setup is shown in Figure 7–20. In general, intermodulation distortion and harmonic distortion will have comparable values for a particular amplifier. However, there is usually some "spread" between the two values. As an illustration, intermodulation distortion could be somewhat higher than harmonic distortion at low test frequencies, but lower than harmonic distortion at high test frequencies. Some variation often occurs between the percentage of harmonic distortion at a low test frequency and a high test frequency. Both harmonic distortion and intermodulation distortion percentages increase rapidly as the high-frequency cutoff point of an amplifier is approached. A similar increase occurs as the low-frequency cutoff point is approached in RC-coupled amplifiers. This increase in distortion results from phase shift through the amplifier circuitry, which becomes greater in the vicinity of cutoff. Normal negative-feedback action in an amplifier is thereby disturbed and the feedback phase trends to positive; in turn, negative-feedback action is partially defeated and the percentage of distortion increases.

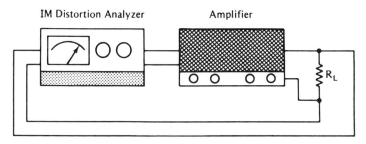

Figure 7–20   Intermodulation distortion test setup.

8

Sophisticated
Recording Techniques

GENERAL PRINCIPLES

Sophisticated recording techniques similar to those used in professional sound studios may be applied to advanced stereo systems. Because high-fidelity microphones have a comparatively low output level, in the range of 1 to 2 millivolts (mV), they are susceptible to hum and noise pickup, unless input circuits follow good practices. Most high-quality microphones operate with balanced lines, as shown in Figure 8–1, that are always shielded. Most professional types of microphones have comparatively low output impedances, such as 150 ohms. This type of microphone can be used with a comparatively long line between the microphone and a recorder. Note, however, that an excessively long line introduces substantial distributed capacitance that acts as a low-pass filter and imposes a treble cut on the audio signal. If a microphone with a balanced line must be connected to a tape recorder that has unbalanced input facilities, an adapter (balun) is required; this adapter transforms the balanced three-wire microphone line to an unbalanced two-wire recorder input circuit.

Nearly all advanced-stereo microphones are of the dynamic design; a diaphragm of thin plastic or metal is attached to a small coil that moves in a magnetic field. A small voltage induced in the coil is conducted by a shielded cable to the tape recorder. Some high-fidelity microphones are of the condenser design, which consists of a capacitor arrangement in which one electrode is a thin diaphragm. The electrodes are charged by DC voltage; as the diaphragm vibrates, the varying spacing between the electrodes generates a small output voltage. This output voltage is so small that a microphone amplifier must be provided

174

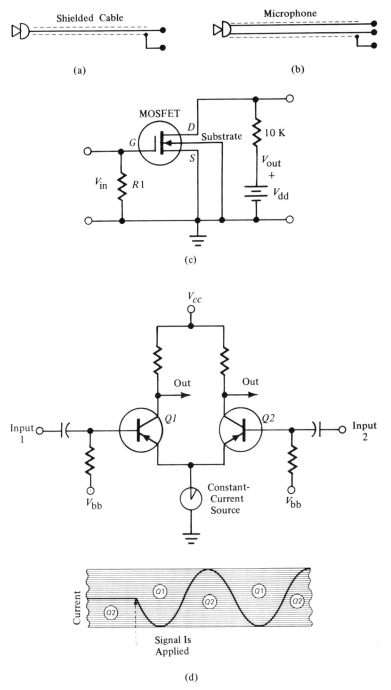

Figure 8–1 Microphone input circuits. **(a)** Unbalanced microphone line; **(b)** balanced microphone line; **(c)** unbalanced recorder input circuit; **(d)** balanced recorder input circuit.

in the same housing. Condenser microphones are capable of extremely high frequency response. A condenser microphone has a very high output impedance and cannot drive an output cable directly. The microphone amplifier not only steps up the small source voltage but also provides a low output impedance for the microphone/amplifier unit.

Most dynamic microphones have a "proximity" effect, which is the result of a bass boost in output from the microphone when a vocalist sings directly into it. Accordingly, many dynamic microphones provide a bass-cut switch, as exemplified in Figure 8–2. When the bass-cut switch is on, an inductor is shunted across the microphone pickup coil.

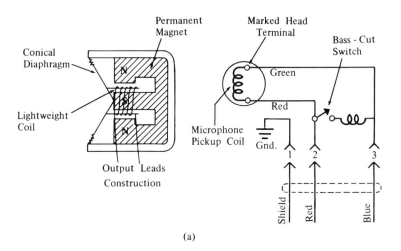

(a)

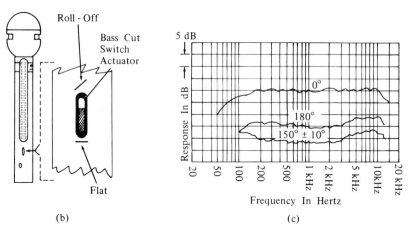

(b)　　　　　　　(c)

Figure 8–2 Microphone circuitry. (a) Configuration for a typical dynamic microphone; (b) bass-cut switch facility; (c) typical frequency response.

Since an inductor has decreasing reactance at lower frequencies, the "proximity" coil introduces a bass cut in the microphone output. A typical bass-cut arrangement provides a roll-off of 6 dB in the range from 1 kHz to 100 Hz.

RIBBON MICROPHONE

Ribbon microphones, a subclass of dynamic microphones, employ a very thin and narrow aluminum ribbon instead of a pickup coil, as shown in Figure 8–3. Note that a ribbon or velocity microphone has a bidirectional (Figure 8–3(a)) acoustic acceptance pattern, whereas a coil-type dynamic microphone has an omnidirectional acceptance pattern. Thus, a ribbon microphone provides maximum response from front and rear, but has zero response to sound waves arriving from either side. Advanced stereo microphones commonly contain combined bidirectional and omnidirectional units that produce a unidirectional

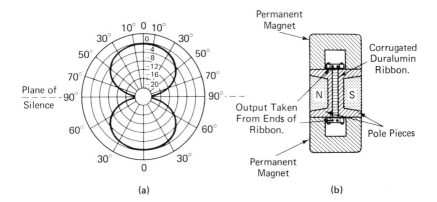

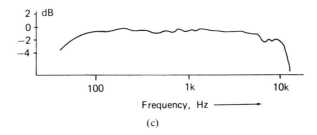

Figure 8–3 Ribbon (velocity) microphone. **(a)** Acceptance pattern; **(b)** basic structure; **(c)** typical frequency response.

cardioid acceptance pattern, as depicted in Figure 8–4. Note that a sound wavefront entering both the back and the front of a ribbon microphone will cancel itself to the extent that its front and back components have the same intensities. In the cardioid design, the front lobe of the bidirectional pattern aids the output from the omnidirectional microphone, whereas the rear lobe of the bidirectional pattern opposes the output from the omnidirectional microphone.

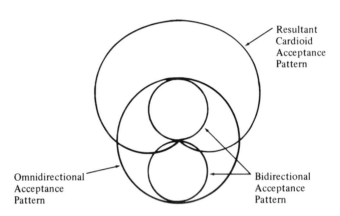

Figure 8–4 Combination of omnidirectional and bidirectional acceptance patterns produces a cardioid acceptance pattern.

CARDIOID ACCEPTANCE PATTERN

The basic cardioid acceptance pattern is also referred to a heart-shaped unidirectional pattern, and its useful acceptance angle is approximately 180 degrees. A cardioid microphone might be described as a "live-front/dead-back" type of transducer. A cardioid acceptance pattern can help to discriminate against noise in recording situations. For example, the microphone can be oriented with its "live side" toward a sound source, and with its "dead side" toward an ambient noise source. Cardioid microphones are widely used during stereo recording. Individual cardioid microphones placed several feet apart are commonly used for recording L and R stereo channels, as depicted in Figure 8–5(a). However, stereophiles sometimes prefer two cardioid microphones oriented to each other at a 90-degree angle and contained in the same housing, as shown in Figure 8–5(b). This latter technique is based on characteristics of the human hearing process, whereas the spaced-microphone technique is less analogous.

Observe the individual acceptance patterns for a dual-cardioid microphone, exemplified in Figure 8–6. The microphones are oriented

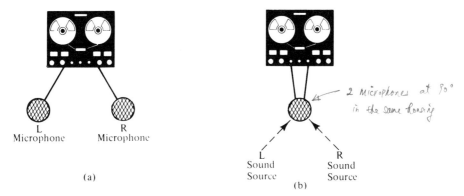

Figure 8–5   Stereo recording. **(a)** Two spaced microphones; **(b)** dual stereo microphone.

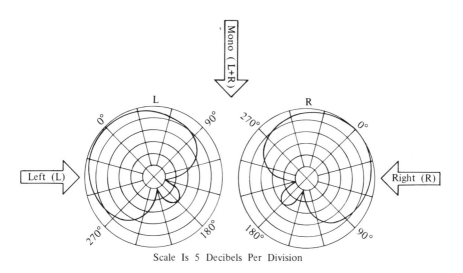

Figure 8–6   Individual acceptance patterns for a dual-cardioid stereo microphone.

to each other at a 90–degree angle. A sound wavefront from the front of the microphone energizes both transducers equally; each microphone reproduces this monophonic L + R signal with only 1 dB loss. On the other hand, a sound wavefront arriving from the left of the microphone is reproduced with 20 dB loss by the R transducer, but with only a 2–dB loss by the L transducer. Therefore, the output from the L microphone is an L signal. Similarly, the output from the R microphone is an R signal. The output from the L and R microphones connected in series is a monophonic L + R signal. Thus, a dual-cardioid stereo microphone operates in almost the same manner as two spaced microphones, and theoretically has the advantage that a dual-cardioid transducer "hears" in the same manner as a human listener. A typical dual-cardioid microphone is illustrated in Figure 8–7.

Figure 8–7   A dual-cardioid stereo microphone, with the L and R microphones contained in the same housing. *(Courtesy, Radio Shack)*

ORDERS OF ACCEPTANCE PATTERNS

Various microphones are designed for different orders of directional acceptance patterns. Thus, four standard orders of acceptance patterns exist for unidirectional microphones, as seen in Figure 8–8. Similarly, there are four standard orders of acceptance patterns for bidirectional microphones, as shown in Figure 8–9. Stereo recording requires a matched pair of microphones, whether they are spaced as individual transducers, or contained in a common housing. As indicated in Figure 8–10, a single cardioid microphone is suitable for recording a piano solo; however, a wide-angle or omnidirectional microphone is preferred

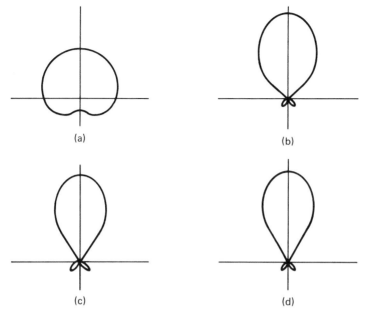

Figure 8–8 Orders of acceptance patterns for unidirectional microphones. **(a)** First order; **(b)** second order; **(c)** third order; **(d)** fourth order.

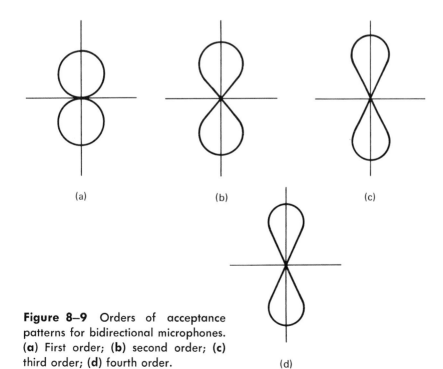

Figure 8–9 Orders of acceptance patterns for bidirectional microphones. **(a)** First order; **(b)** second order; **(c)** third order; **(d)** fourth order.

for a vocalist with piano accompaniment; conversely, a narrow-angle microphone helps to reduce ambient noise for a vocalist with self-accompaniment. A vocalist with accompanying ensemble may prefer to sing into a highly directional microphone.

When a vocalist is accompanied by an ensemble, as exemplified in Figure 8–10(d), the vocalist's microphone will not necessarily match

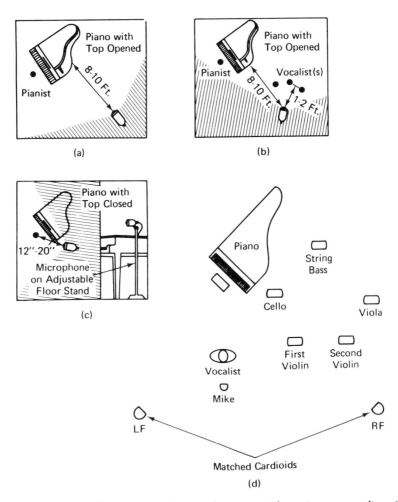

Figure 8–10 Different types of microphones provide optimum recordings in various situations. **(a)** Cardioid microphone is suitable for recording a piano solo; **(b)** wide-angle (or omnidirectional) microphone is preferred for a vocalist with piano accompaniment; **(c)** narrow-angle microphone reduces ambient noise for vocalist with self-accompaniment; **(d)** vocalist with accompanying ensemble may prefer a highly directional microphone.

the LF and RF microphones. In this situation, the vocalist may prefer a highly directional microphone in order to subdue both ambient noise and loud instrumental passages, but use cardioid LF and RF microphones. A similar application is pictured in Figure 8–11. Advanced stereo systems balance the output levels from the vocalist's microphone and from the L and R microphones with an audio mixer; volume controls on the mixer are adjusted as required to provide an optimum blend and balance for the recorder. If a formal group's performance is to be recorded, the vocalist's microphone may be mounted on a floor stand placed several feet in front of the vocalist. Note that hand-held microphones should always be provided with wind screens in order to minimize "popping" and "breathing" noises. These extraneous noises are less evident when a microphone is placed several feet in front of the vocalist.

Figure 8–11 A vocalist usually prefers to sing directly into a highly directional microphone.

When a dialogue is to be recorded, a bidirectional microphone is most appropriate, as depicted in Figure 8–12. Its bidirectional acceptance pattern provides full response to speech wavefronts from the participants, whereas ambient noises from either side of the microphone are largely rejected. Although a fourth-order acceptance pattern discriminates against ambient noise to a greater extent than does a first-order acceptance pattern, the former requires closer attention to recording procedure. Consider a dialogue recording, such as the one depicted in Figure 8–12. If a bidirectional microphone with a fourth-order acceptance pattern is used, the participants must not move from their initial positions. If a participant leans to one side to reach for notes, an ash tray, or a glass of water, his voice level will be attenuated accordingly, and this portion of his conversation may be lost in the recording.

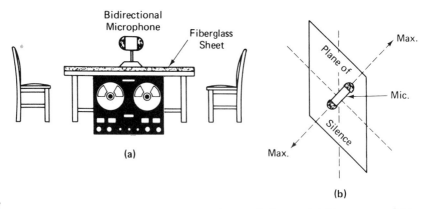

Figure 8–12   Arrangement for recording a dialogue. **(a)** Placement of bidirectional microphone; **(b)** plane of silence is at right angles to microphone.

MISCELLANEOUS APPLICATIONS

When an extremely directional microphone is required, the transducer mounted at the focus of a large parabolic "dish" can focus the incident sound wavefront. This technique is very useful when a particular weak sound source is to be "picked out" from adjacent sources, and when the desired source is quite far from the microphone. The main disadvantage of this method is that the parabolic "dish" occupies a large area. The diameter of the parabola must be equal to approximately one wavelength of the lowest sound frequency that is to be recorded. Note that a parabolic sound reflector cannot discriminate between a desired sound wavefront and ambient noise. Thus, the recorded sound may have an unexpectedly high background noise level. Because of the large area required by a parabolic "dish", some audio aficionados prefer the "short gun" type of directional microphone. This type of transducer originally resembled a machine gun; it consisted of a circular bundle of tubes of widely varying lengths. Because of its overall length, the original design was superseded by the "short gun" directional microphone, which employs a much shorter circular bundle of tubes. These tubes are perforated in patterns that produce phase shifts at various frequencies for sound waves arriving from the side of the microphone, and thereby optimize the highly directional characteristic of the microphone.

If a unidirectional microphone is to be used for recording a dialogue between a tall person and a short person, as illustrated in Figure 8–13, the tall person should be seated while the short person stands. Thus, with both sound sources at approximately the same level, a balanced recording is more likely to be obtained. However, if objectionable unbalance remains, one person or the other may be moved nearer to or farther away from the microphone. The resourceful operator may also introduce one or more acoustic screens both to improve the balance of the recording, and to effectively control the reverberation time of the recording acoustic environment. Thus, if the performers are partially "boxed in" by highly reflective acoustic screens, the recorded material will be quite "lively." On the other hand, if highly absorbent acoustic screens are used, the recorded material will be "dead."

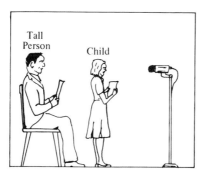

Figure 8–13 Obtaining a balanced recording with a unidirectional microphone.

MICROPHONE FREQUENCY RESPONSE

A microphone that is suitable for advanced-stereo applications should have a frequency response that is flat within ± 1 dB from 20 Hz to 20 khz. Only a few designs can meet this requirement. Many good-quality microphones have frequency responses that are flat within ± 3B or ± 4 dB over their frequency ranges (typically up to 10 kHz). Note that only a few designs provide high frequency response in excess of 12 khz; many good-quality microphones exhibit substantial peaks and valleys at frequencies greater than 10 kHz. Many microphone designs provide good low-frequency response, but some will be found deficient in bass response. An exceptional microphone design that has high-frequency response well above 12 kHz is likely to exhibit a very narrow high-frequency lobe and a broad low-frequency lobe. In high-fidelity applications, such a microphone must be oriented with its axis directly toward the sound source.

Although a mediocre cardioid microphone may exhibit a comparatively narrow treble lobe, as exemplified in Figure 8–14, this lobe can be widened by introducing a pair of acoustic screens. Treble screens are relatively small in area; they have a high reflectance factor for high audio frequencies with a negligible reflectance factor for low audio frequencies. In other words, an acoustic screen does not have appreciable reflectance for wavelengths that are longer than the length of the screen. Redirection of treble sound rays occurs as shown in Figure 8–15. If two pairs of screens are utilized, the effective width of the treble lobe can be further increased. The additional screens are placed quite far apart and farther back from the microphone. The chief precaution in screen placement is to avoid parallelism between any two surfaces. Parallel screens tend to develop standing waves, which cause resonance and excessive reverberation.

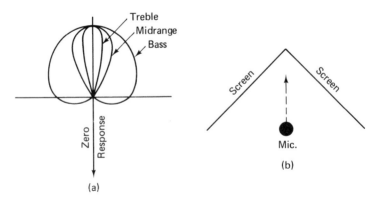

(a)

(b)

Figure 8–14 A mediocre cardioid microphone has a comparatively narrow treble lobe. (a) Frequency patterns; (b) acoustic screens effectively widen the treble lobe.

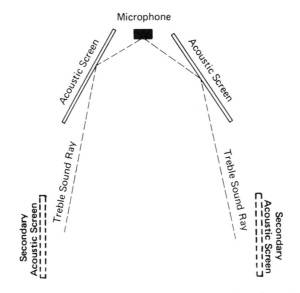

Figure 8–15 Small reflective screens effectively widen the treble acceptance pattern of a microphone.

CONTACT MICROPHONES

Electrical musical instruments such as the electrical guitar depicted in Figure 8–16 usually employ contact microphones. The microphone output is fed to an audio amplifier (occasionally via a mixer) and thence

Figure 8–16 An electrical guitar is usually recorded with contact microphones.

to a speaker system or tape machine. Volume controls and tone controls common on electrical musical instruments allow the performer to "tailor" the sound output. Note that musical instrument amplifiers are not necessarily high-fidelity designs. For example, a typical musical instrument amplifier is rated for a total harmonic distortion (THD) value of 5 percent. Moreover, musical instrument speakers are not necessarily high-fidelity units. Better recordings can be obtained if the output from a contact microphone is fed directly to the preamplifier in a tape machine, instead of using a separate microphone to pick up the sound radiated from the musical instrument speaker. Of course, the musical instrument speaker should continue to operate during the recording, so that the performer can remain in his normal acoustic environment.

RECORDING OF RADIO OR TV SOUND

Radio or TV sound may be recorded by placing either an acoustic link or with an electrical connection between the receiver and the tape machine, as depicted in Figure 8–17. Note that if the sound radiated by the speaker is picked up by a microphone to energize the tape machine, the recorded material will be colored considerably by the

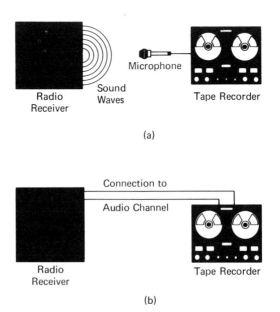

(a)

(b)

Figure 8–17   Recording of radio program. **(a)** Recording will include coloration of listening area, speaker distortion, and microphone distortion; **(b)** preferred method uses audio circuit connection between radio and recorder.

characteristics of the speaker, by the acoustics of the recording room, and to some extent by the characteristics of the microphone. On the other hand, if the output from the receiver is connected directly to the preamplifier in the tape machine, these sources of distortion are eliminated. If the tape machine is connected at the output of the de-modulator in the receiver, distortion can also be eliminated from the receiver's audio-output stage. Note that if the output from a small AC-operated radio is to be recorded, an extremely high hum level may occur, owing to "hot chassis" construction in the receiver. In such a case, a line-isolation transformer must be used to power the receiver.

9

Public-Address Sound Equipment

GENERAL REQUIREMENTS

Various types of public-address (PA) sound equipment in general use are illustrated in Figure 9–1. Both indoor and outdoor systems are available; outdoor systems have weatherproof folded-horn speakers. Many wide-coverage (high-power) systems employ sound columns that typically contain six cone-type speakers in each column. A typical PA amplifier contains five mixable inputs that can accommodate four microphones and an auxiliary/phono source. Each mixer input has a calibrated volume control. Master volume and tone controls are included for convenient variation of the total sound-output level. More elaborate designs have a priority paging switch for one of the microphone inputs that permits the operator to "break in" for special announcements. An output jack may also be included for tape-recording the proceedings. Many designs feature constant-voltage output circuitry that allows speakers to be added or removed without consideration of impedance-matching requirements. A versatile PA amplifier is illustrated in Figure 9–2.

PA systems have to operate in a wide variety of acoustic environments. For instance, in a convention hall a PA system operates in an enclosed space, although the environment may be very noisy at times. On the other hand, at an athletic event a PA system operates in an open area in which the noise level may be high or low. Elsewhere, a PA system may operate in a very noisy aural environment, as in a factory (Table 9–1). Conversely, a PA system in a hotel lobby usually operates in a quiet environment. Acoustics are sometimes a troublesome factor, as in a large railway depot or a large airport lobby with many corridors

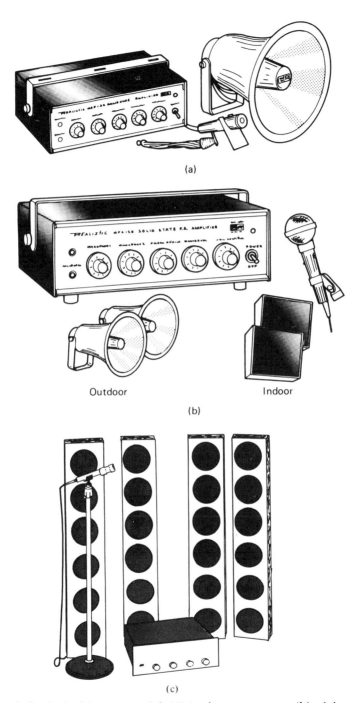

(a)

Outdoor   Indoor

(b)

(c)

Figure 9–1   Basic PA systems. **(a)** Minimal arrangement; **(b)** elaborated arrangement with indoor and outdoor speakers; **(c)** wide-coverage system with sound columns. *(Courtesy, Radio Shack)*

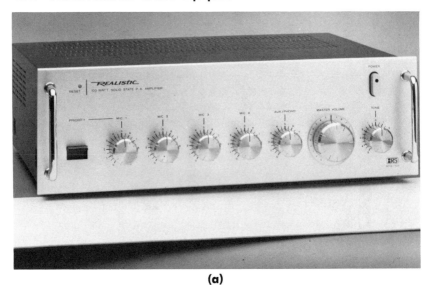

(a)

(b)

Figure 9–2 A versatile 100-watt PA amplifier and typical outdoor speaker. (a) View of control panel; (b) weatherproof folded horn. *(Courtesy, Radio Shack)*

TABLE 9–1

SOUND LEVEL CHART

General Location Description	Noise Level In dB	Noise Quality
Factory (Very Noisy)		Very Noisy—
Machine Shop (Average)	90	Conversation
Printing Plant	80	Difficult or Impossible
Ball Room (Normal Use)		Noisy—
Restaurant (Noisy)		Voice Must Be
Assembly Line (Noisy)	75	Raised To Be
Factory (Average)		Understood
Machine Accounting Area	70	
R.R. or Bus Depot		
Auditorium (Average)		
Shipping/Receiving Dept.		
Department Store		Normal—
Auditorium (Quiet)	65	Normal
Restaurant (Average)		Conversation
Store (Quiet)	60	Easily
Office (Quiet)		Understood
Hotel Lobby	55	
Doctors Waiting Room		

Courtesy, Allied Electronics

where the environment is highly reverberant with lingering echoes. In this situation, numerous small speakers can be spaced advantageously at close intervals. When the noise level is high, more audio power must be applied to the speakers than when the noise level is low (Table 9–2).

BASIC SPEECH REINFORCEMENT

Since the human voice has a low energy level, it can ordinarily cover only a limited area, even in a good acoustic environment. For this reason, speech reinforcement systems are in wide use. Few such installations qualify as high-fidelity sound systems, however, as some of them reproduce very mediocre sound. The chief function of a speech-reinforcement system is information transfer, or intelligibility. If the spoken words can be properly understood, even a mediocre reproduction of a speaker's voice at a distance of a hundred feet from its source

TABLE 9–2

SPEAKER INPUT POWER CHART

The chart below has been designed to help you determine your speaker and
amplification requirements in relation to the environmental noise level.

Total Number of Speakers	Total Amplification Recommended (in watts)		
	Normal	Noisy	Very Noisy
1	½	2	5
2	1	4	10
3	1½	6	15
4	2	8	20
5	2½	10	25
10	5	20	50
20	10	40	100
50	25	100	250
100	50	200	500
150	75	300	750
200	100	400	1000
500	250	1000	2500
750	375	1500	3750
1000	500	2000	5000

Courtesy, Allied Electronics

serves a useful purpose. A wide frequency range is not required for
practical articulation in a speech-reinforcement system, as indicated
in Figure 9–3. A frequency range of 200 Hz to 2 kHz is generally
considered adequate for satisfactory articulation. However, this limited
frequency range impairs the "naturalness" of the speaker's voice to
some extent.

Because a speech-reinforcement system is primarily a means of
information transfer and not a mode of entertainment, two sound
sources must be considered, as illustrated in Figure 9–4. A listener
positioned half way between a podium and a speech-reinforcement
speaker will hear almost the same sound from the podium and from
the speaker. In other words, part of the perceived sound energy arrives
from the orator, and another part of the sound energy arrives from
the speech-reinforcement speaker. Since both sound wavefronts are
perceived simultaneously by the listener, there is no delay distortion
present to impair articulation. Only the total sound level is increased
in this situation. Of course, this situation would be modified if the
acoustic environment were not suitable for reproduction of a second
source to the rear of the listener.

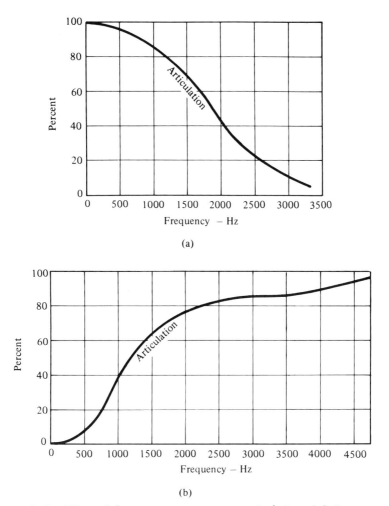

(a)

(b)

Figure 9–3 Effect of frequency response on articulation. **(a)** Percentage articulation when low frequencies to left of curve are cut off; **(b)** percentage articulation when high frequencies to right of curve are cut off.

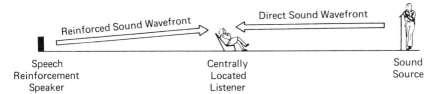

Speech Reinforcement Speaker

Centrally Located Listener

Sound Source

Figure 9–4 Centrally located listener perceives a higher total sound level than from front or rear alone.

Consider next a listener seated toward the rear of a highly reverberant hall, as depicted in Figure 9–5(a). Sound is substantially reflected from the far end of the hall, even though no speech reinforcement equipment is used. The listener experiences an increased sound level owing to the reverberant energy. On the other hand, the reverberant energy arrives after an excessive time delay and creates an area of confusion and poor articulation. Another situation that results in an area of confusion is shown in Figure 9–5(b). Here, the listener is seated toward the rear of a long hall that has acceptable acoustic characteristics and that contains a speech-reinforcement speaker at the

Long Highly Reverberant Hall

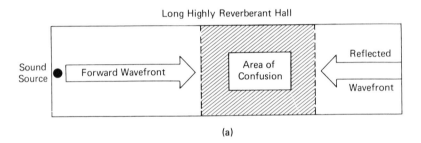

(a)

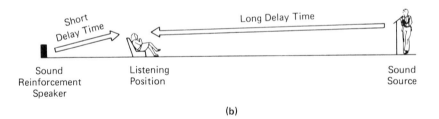

(b)

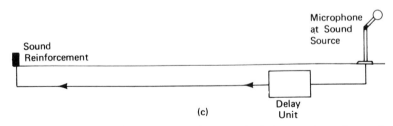

(c)

Figure 9–5 Excessive delay time reduces articulation and confuses the listener. (a) Long, highly reverberant hall; (b) rear sound-reinforcement speaker improves articulation; (c) delay unit in audio line improves articulation at rear of hall.

rear. As before, the listener finds himself in an area of confusion owing to the disparity in arrival times between the sound-reinforcement wavefront and the direct sound wavefront. This situation can be corrected by a time-delay unit in the audio line, as shown in Figure 9–5(c), which will equalize the time of arrival of the sound-reinforcement wavefront and the direct sound wavefront at the listening position.

PRINCIPLES OF AUDIO TIME DELAY

Excessive disparity in arrival times of two sound wavefronts carrying the same information results in poor intelligibility, as noted above. The maximum tolerable time delay between direct and reinforcement wavefronts is approximately 25 milliseconds, which varies to some extent depending upon the relative intensities of the two sound wavefronts. A time delay (echo time) greater than the critical value results in a confusing "separation effect" that impairs intelligibility. The listener's direction-perception or position-perception process jumps from one source to the other, and he becomes distracted or annoyed. In addition, syllabic overlap impairs articulation. Inasmuch as the human ear has a natural tolerance for everyday echo effects, a listener will tend to disregard an echo, even in mediocre acoustic environments. However, when an echo is excessively delayed, the listener can no longer blend the two wavefronts and he experiences a feeling of confusion. He reacts with annoyance to the split-sound sources. (See Figure 9–6).

The critical value in wavefront arrival times depends somewhat upon the relative sound intensities of the wavefronts. Under ordinary circumstances, an echo is weaker than the direct wavefront from the sound source. It is understandable that the human ear has less tolerance for echoes that have greater intensities than the direct wavefront from the source. For example, in a sound-reinforcement system, 10 percent of the listeners will be disturbed if the PA sound level is 3 dB below the source level and is delayed 60 milliseconds. Again, 10 percent of the listeners will be disturbed if the PA sound level is 6 dB below the source level and is delayed 80 milliseconds. On the other hand, 10 percent of the listeners will be disturbed if the PA sound level is 10 dB above the source level and is delayed 30 milliseconds. Therefore, no listener within range of the PA sound field should be exposed to two sound wavefronts that are separated more than 25 milliseconds in time. This requirement is summarized in the Haas-effect diagram shown in Figure 9–7.

Figure 9–8 indicates that for echoes weaker than the direct sound wavefront echo delay time is approximately in direct proportion to

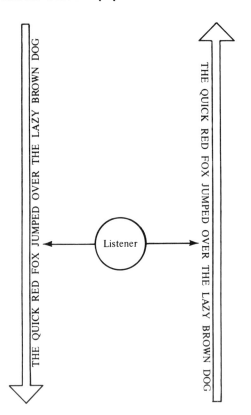

Figure 9–6   Visualization of split-sound source perception.

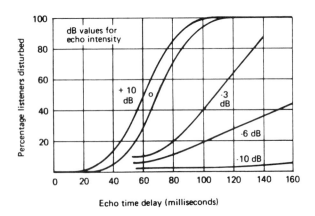

Figure 9–7   Time-delay disturbance versus intensity of a reflected sound.

echo intensity. On the other hand, this relationship increases out of proportion with great rapidity as the echo intensity begins to exceed the intensity of the direct sound wavefront. Thus, if the guideline of less than 10 percent listener disturbance is observed, a delay time of 35 milliseconds and an echo intensity of +10 dB represent the absolute maxima for a sound-reinforcement system. Conservatively, a delay time of 25 milliseconds and an echo intensity of +6 dB would be considered proper design limits. Note that the comparative intensities of a direct wavefront and a sound-reinforcement wavefront can be measured with a sound-level meter by first switching out the direct speaker(s), and then by switching out the sound-reinforcement speaker(s).

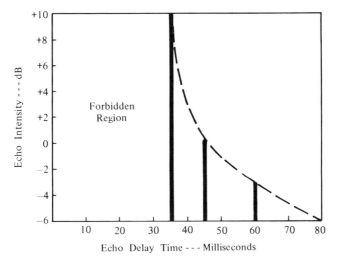

Figure 9–8 Maximum permissible echo delay time versus intensity for less than 10 percent listener disturbance.

DELAY COMPENSATION

Although a cluster of speakers (Figure 9–9), or a speaker column can provide considerable acoustic coverage, the substantial reverberation characteristic of large listening areas may be problematic. In this case, audio systems can be improved by supplementing a large speaker cluster or column with smaller speaker installations located back from the primary sound source at successive intervals. In turn, the primary

Figure 9–9   A speaker cluster for an open-air PA system.

source can be operated at a lower sound level; its output is progressively reinforced by the smaller speaker installations. In this way, all of the listeners experience a more direct sound and less reverberant sound. As an illustration, a large speaker cluster or column installed near the rostrum may cover a distance up to 100 feet. Beyond this point, additional coverage may be provided by a supplementary speaker column, as depicted in Figure 9–10.

At a distance of 100 feet, an acoustic delay device is needed to process the audio signal and to make the sound from the first column arrive 5 milliseconds before the sound from the second column is radiated. Next, at the following 50–foot interval, a second supplementary column may be installed with a time-delay device that allows

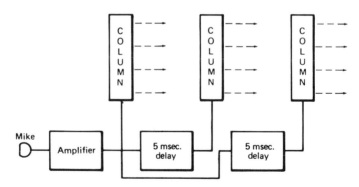

Figure 9–10   PA system with sound columns driven through acoustic delay units.

associated listeners to perceive a sound wavefront from the main column first, followed 5 milliseconds later by the wavefront from the first supplementary column, and then after another 5–millisecond delay followed by the sound wavefront from the second supplementary column. Note also that the intensity of the delayed sound should not be more than 10 dB greater than the intensity of the wavefront that arrives first. Otherwise, the listener is likely to experience a split-sound effect ("acoustic separation") of the two sources.

Many time-delay units for PA systems employ a tape deck having suitably spaced playback and recording heads, as shown in general outline in Figure 9–11. "Bucket-brigade" large-scale integrated (LSI) delay devices or magnetic disks can also be used, but it is essential to limit the amount of sound energy that is radiated from a speaker toward the front of the building or listening area (see Figure 9–10). In other words, rearward radiation from a delayed sound source should be at least 6 dB less than its forward radiation level. A helpful technique is portrayed in Figure 9–12. A layer of cotton 2 inches thick is applied to the back of the speaker·column. This padding not only absorbs both incident and diffused sound energy, but also changes the phase of the sound that is radiated from the back of the speaker by an amount equal to the phase change that occurs when the sound waves travel from the front of the column around to its back. These acoustic reactions appreciably reduce the intensity of the rearward-radiated sound. In particularly difficult situations, the ill effects of rearward radiation can be reduced by restricting the frequency response of the PA system to a range of 250 Hz to 4 kHz.

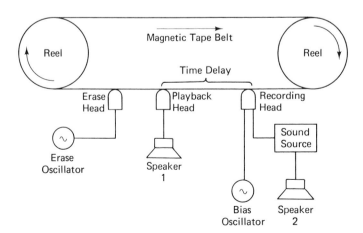

Figure 9–11 Principle of a magnetic-tape audio delay unit.

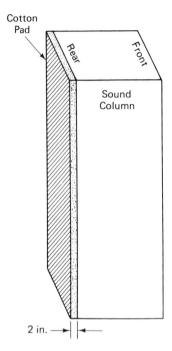

Figure 9–12 Cotton pad reduces rearward acoustic radiation from sound column.

EQUALIZATION AND FEEDBACK

If positive acoustic feedback from speaker(s) to microphone exceeds a critical value, the PA system will "howl" at the frequency of maximum feedback. This principle is demonstrated in Figure 9–13(a). Excessive feedback can be controlled in several ways. The frequency range of the system can be reduced to 200–2000 Hz. This is the minimum frequency range for acceptable articulation. Another approach is to use a "short-gun" microphone, as depicted in Figure 9–13(b). This design is highly directional at frequencies above 500 Hz, and has a cardioid acceptance pattern for frequencies below 500 Hz. This type of microphone can be oriented to discriminate against positive acoustic pickup and thereby to suppress "howling." In other situations, a frequency equalizer such as the one pictured in Figure 9–14 can be used to restrict the system frequency response as required, and/or to "cut a dip" into the frequency characteristic at a suitable point. Positive acoustic feedback is usually most troublesome in the midrange region, and if

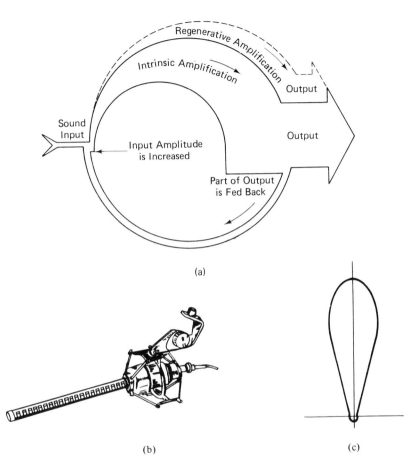

(a)

(b)

(c)

Figure 9–13 Principle of positive acoustic feedback. **(a)** Feedback energy adds to input; **(b)** "short-gun" type of microphone; **(c)** acceptance pattern.

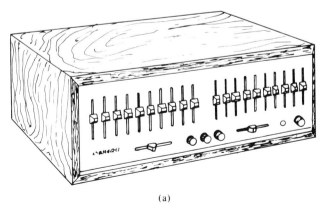

(a)

Figure 9–14 Frequency equalizer. **(a)** Appearance;

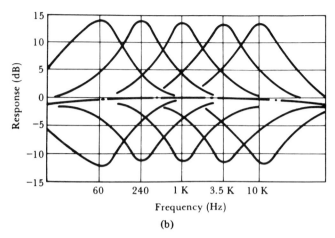

Figure 9–14 *Continued:* **(b)** adjustable frequency responses. *(Courtesy, Dynaco)*

the system gain is suitably reduced in a sector of this region, "howling" can generally be controlled without an objectionable tradeoff in articulation.

AUDIO MIXER

If a PA amplifier does not have mixing facilities, a separate audio mixer unit can be used, as shown in Figure 9–15. This unit has four variable-level inputs with extensive isolation. A single output terminal is provided for connection to the input of the PA preamplifier. Emitter-follower output is employed in the mixer unit to accommodate a wide range of preamplifier input impedances. The field-effect transistor provides a gain of 17 dB, so that a wide range of input signal levels can be effectively mixed. Because the bipolar transistor substantially increases current without increasing voltage, it is useful when a PA amplifier with moderate or low input impedance is to be driven by the mixer.

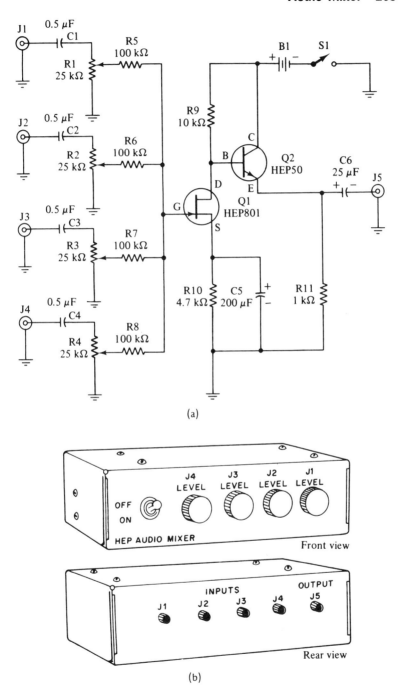

(a)

(b)

Figure 9–15 Audio mixer. **(a)** Configuration; **(b)** appearance. *(Courtesy, Motorola)*

10

Miscellaneous Audio Equipment

GENERAL SURVEY

A wide variety of miscellaneous audio equipment is used in home entertainment, in commercial communications, in law enforcement activities, in medical care, in scientific investigations, and so on. Intercommunication units installed in residences are often combined with FM/AM radios that may be supplemented with eight-track tape decks and/or high-fidelity record players. Many units are designed as on-wall or in-wall types, as illustrated in Figure 10–1. The master unit with the FM/AM radio (see Figure 10–2) may be installed in the wall, whereas the substations are on-wall units. A substation has only a speaker, whereas a master station includes both a speaker and a microphone. Various switching facilities are available; for example, a master station may be able to call any one substation, or any group of substations. If an intercom system consists entirely of masters, the switching facilities generally permit several separate conversations to be carried on simultaneously without interfering with each other.

SPECIAL INTERCOM ARRANGEMENTS

A combination intercom system is relatively flexible. It includes both master stations and substations. Master stations can talk to each other and to each substation in the system selectively; or, one or more substations can be exclusively switched to only one master station. As well, each master station can be switched for "private" operation, so that other master stations cannot listen to the conversation until the

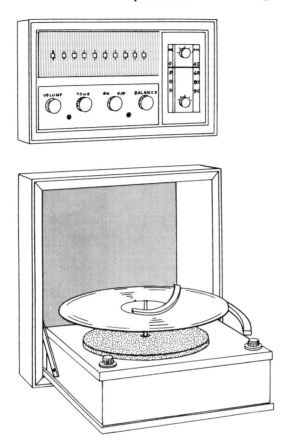

Figure 10–1 In-wall type of intercom unit, with supplementary record player.

initiating master station is switched for "non-private" operation. A high-volume master station can be set for ten times normal volume output, so that the incoming message can be heard over a large area. Although most intercom equipment is used for conversations, some is also used for paging. Substantial audio power is required to page a noisy area. A typical seven-station intercom master station is connected to six remote stations, with one channel used to drive a paging amplifier.

Many intercom systems are interconnected with multiconductor cables. However, wireless intercom systems, depicted in Figure 10–3, are also widely used. A wireless intercom unit is merely plugged into common AC outlets and is ready to operate in the home, office, warehouse, or any area provided with 117–volt service. The only limitation in wireless-intercom operation is that audio signals cannot pass through

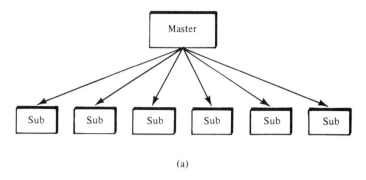

(a)

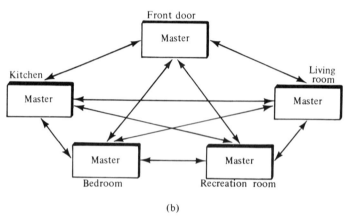

(b)

Figure 10–2 Basic intercom arrangements. **(a)** One master station with several substations; **(b)** two-way system throughout, with master stations only.

Figure 10–3 A two-station wireless intercom assembly. *(Courtesy, Allied Electronics, a Tandy Corporation Company)*

a power transformer. Therefore, all wireless intercom units in a system must be plugged into a common power line. Typical designs provide two channels, and can be expanded with two additional single-station units. Two conversations may be accommodated simultaneously. Carrier frequencies of 225 kHz and 160 kHz are typical, but other ultrasonic frequencies may be used. Both AM and FM designs are available. AM wireless intercom systems tend to develop background noise owing to interference from appliance motors, fluorescent lights, dimmer switches, and so on, whereas FM wireless intercoms have quieter operation.

Specialized intercom units are available for monitoring a baby's room, sick room, or any area for purposes of safety or protection. This type of intercom employs a miniature radio transmitter that operates through any AM radio (even a car radio) at distances up to 300 feet. The intercom is tunable for selection of a desired clear channel in the frequency range of 1200 to 1650 kHz. Basically, this arrangement is a wireless microphone. Another form of wireless microphone, depicted in Figure 10–4, is an FM unit that comprises a high-fidelity microphone and a miniature FM transmitter that operates through an FM radio receiver at a distance up to 300 feet. This form of wireless microphone is used on the stage, in the home, or in industrial areas. It operates from a self-contained 9–volt battery, and has an auxiliary input connector as well as output facilities for a tape recorder.

Figure 10–4 A wireless FM microphone for intercom or professional applications. *(Courtesy, Allied Electronics, a Tandy Corporation Company)*

TELEPHONE ANSWERING EQUIPMENT

Various types of telephone answering equipment are in wide use, by commercial establishments, professional offices, and private residences. These devices are often termed telephone valets or automatic valets. A typical design, such as the one pictured in Figure 10–5, provides automatic answering and message-taking service. An induction pickup unit is used; connection to the telephone line is made with a four-pin phone plug. When the telephone rings, the automatic valet answers and delivers a prerecorded message; it records callers' messages on

Figure 10–5 An automatic valet records telephone calls and transmits prepared messages. *(Courtesy, Newark Electronics)*

any standard tape recorder with remote control facilities. Some designs feature built-in cassette recording facilities, and other types permit the user to listen to incoming calls without touching the telephone, and thus enable him to "screen out" unwanted calls. One design provides a ring-adjust control, that can set the valet to answer after the first, second, or third ring, etc.

A remote-mate unit for a phone valet permits the user to play back recorded messages from any public or private telephone. He calls his own number from any telephone, presses the button on his hand-held phone activator, and the valet responds by automatically resetting itself after playing back any messages that have been recorded. For example, the user may be away from his phone and expecting an important message. When he calls his phone number, he holds the activator up to the phone from which he is calling, and presses the "on" button. This causes a specially encoded signal to activate the valet at his office and to play back any recorded messages. Each remote-mate unit has a unique encoded signal that prevents others from gaining illegal access to the valet. The activator is battery-powered with flashlight cells.

TELEPHONE DIALERS AND AMPLIFIERS

An automatic telephone dialer and amplifier is depicted in Figure 10–6. It can be programmed with twenty-six telephone numbers; a name panel with a selector knob accommodates up to twenty-six names. When the selector knob is turned to a particular name and the call button is pressed, the automatic dialer is activated. The program can be erased and other numbers entered into the memory of the unit. To reprogram the memory, a desired number is dialed once; thereafter, this number will be automatically dialed whenever the command is given and the call button

Figure 10–6 An automatic telephone dialer/amplifier arrangement. *(Courtesy, Allied Electronics, a Tandy Corporation Company)*

depressed. The telephone amplifier pictured in Figure 10–6 allows a group to monitor an incoming call. If the dialer-amplifier is switched out, the telephone operates in the conventional manner. Connection to the telephone line is made with a four-pin phone plug. A miniaturized snap-on telephone amplifier and hearing aid is pictured in Figure 10–7. This design is used by the hard-of-hearing; it contains a volume control and operates from a small battery.

Figure 10–7 A snap-on telephone amplifier and hearing aid. *(Courtesy, Lafayette Electronics)*

An amplifier can be driven with a telephone pickup coil, as depicted in Figure 10–8. It is an electromagnetic arrangement with a suction cup for attachment to the housing of the telephone receiver. A pickup coil provides operating convenience because a plug is not required and no connection is made to the telephone line. Note that telephone amplifiers may also be driven by means of acoustic coupling, without any connection to the telephone line. In other words, the receiver is placed over a microphone arrangement that generates a corresponding audio signal to energize the amplifier. For example, acoustic coupling is employed in the design pictured in Figure 10–9. When the phone rings, the handset is then placed in the cradle of the telephone

Figure 10–8 A telephone pickup coil. *(Courtesy, Lafayette Electronics)*

Figure 10–9 An acoustically coupled telephone amplifier. *(Courtesy, Newark Electronics)*

amplifier, and the incoming message is reproduced by the loudspeaker. This simple arrangement permits use of the telephone while the hands are free to type, check orders, take notes, and so on. The device operates from a 9-volt transistor battery.

SURVEILLANCE AND LAW ENFORCEMENT EQUIPMENT

Specialized audio equipment is used in surveillance and law-enforcement operations. One familiar unit is the electronic megaphone or "bull-horn" pictured in Figure 10–10. It consists of a weatherproof folded-horn speaker with a cardioid microphone and audio amplifier. The amplifier provides approximately 15 watts of audio power to the speaker, and is powered by self-contained batteries. When operated at maximum volume, the electronic megaphone is audible for more than one-half mile. Because the cardioid microphone has practically zero response to sound waves arriving from the rear, objectionable acoustic feedback can be avoided. An electronic megaphone can also be operated from an auxiliary microphone and extension cable. A directional microphone should be used in this mode of operation and oriented as required to eliminate excessive acoustic feedback and resultant "howling."

Figure 10–10   An electronic microphone with built-in 15-watt amplifier. *(Courtesy, Allied Electronics, a Tandy Corporation Company)*

Various types of listening devices ("bugs") are used in surveillance activities. As an illustration, a microphone with a suction-cup form of housing is employed for listening to or recording conversations through doors, windows, or floors. The suction cup holds the microphone in firm contact with the surface, and converts its vibrations into a corresponding audio signal. This technique is analogous to the traditional practice of pressing one's ear to a railroad track to hear a distant train. In some surveillance operations, a suction-cup microphone is connected to a wireless microphone, such as the one pictured in Figure 10–4. In this way, a conversation can be monitored remotely with a radio receiver. An equivalent arrangement is also available in miniaturized form that can be inserted inside a telephone handset. In turn, conversations can be heard at any chosen remote location whenever a person speaks into the "bugged" telephone handset.

Some walls are more soundproof than others. When rigid double walls separate adjoining rooms, an investigator or detective often uses a microphone having an extension pipe approximately 1 foot in length. The small-diameter pipe responds both to sound waves and to wall vibration. In a typical application, the investigator removes an electric outlet box from the wall, so that the extension pipe can be inserted through the wall and held against the wall of the adjoining room. In many cases, the microphone output can be sufficiently amplified to make conversations in the adjoining room easily audible. In difficult situations, the investigator may greatly increase the sound level at the microphone by boring a small hole through the wall to the adjoining room. This technique is analogous to the speaking-tube communication systems that were used in apartment buildings prior to the advent of intercommunication equipment. Some surveillance-type microphones are designed with flexible extension pipes that can be curved as required in order to gain access to an obstructed position.

Audio waveform recorders (XY plotters) are used in interrogation procedures in combination with polygraphs (lie detectors). A deception indicator associates stress conditions with the act of prevarication. Thus, the operator analyzes changes in the subject's blood pressure, respiration rate, skin resistance, pulse rate, speech rate, vocal timbre, and peak sound level. For instance, the operator checks words and phrases in the audio waveform that exhibit a sudden increase in peak level. This increase casts suspicion of prevarication upon the associated word or phrase. An increase in average pitch of the audio waveform is also regarded with suspicion. Such changes in the audio waveform are correlated with the other physiological factors noted above. If the change in audio waveform is accompanied by a sudden rise in blood pressure, respiration irregularity, decrease in skin resistance, and acceleration in pulse rate, the associated word or phrase will be regarded with greater suspicion. Because results of polygraph tests are seldom clear, considerable reliance must be placed on the experience and "track record" of the operator.

MEDICAL AUDIO EQUIPMENT

The audiometer is one of the most familiar units of medical audio equipment. It is an electronic instrument for measuring hearing acuity. Generally, the listener is given an earphone or a headset and is asked to report his perceptions of an audio signal (usually a sine wave) as its intensity and its frequency are varied. The response of the left ear is not necessarily the same as the response of the right ear. Some elaborate audiometers emit a variety of test signals, such as sine-wave tones, white noise, phrases, and nonsense syllables that may be reproduced by earphones, speakers, or bone-conduction devices. These sophisticated test procedures are helpful in diagnosing hearing abnormalities, as well as in prescribing appropriate types of hearing aids. In general, young people have a lower threshold of hearing and they can hear higher audio tones than older people. An audiogram is a graph that shows hearing loss, percent of hearing loss, or percent of hearing as a function of frequency.

The stethoscope is also a familiar medical device. Modern medical technology has supplemented it with the heart-sound amplifier and XY plotter. Thus, a patient's heart, lungs, and other internal organs can be examined on the basis of sound waves that they produce. Amplification permits the sounds to be reproduced by a speaker so that they can be monitored by more than one physician. Amplification also permits XY recordings to be made for subsequent review and analysis. This type of recording is called a phonocardiogram. Computer analysis of waveforms is becoming a trend in this general field, because parameters can be evaluated that would otherwise be difficult to determine.

In a related technique, high-frequency sound waves are applied at some point on the patient's body by means of a transducer, and the resulting reflections and refractions of the sound waves are applied to an oscilloscope or an XY plotter for analysis of modifications.

AUDIO ANALYSIS EQUIPMENT

Audio waveform analysis and research activities make extensive use of the sound spectrograph. This unit separates an audio waveform into twelve frequency bands, from bass through midrange to treble tones. Bandpass filters are used in this separation process. Each filter has a changing output that corresponds at any instant to the amplitude of the audio signal within a particular frequency band. A more elaborate design of the sound spectrograph divides the audio waveform into 500 frequency bands and records band outputs versus time by means of an XY plotter. Another basic type of audio analysis equipment, which is called the speech synthesizer or voder, employs audio oscillators and filters, as well as a keyboard to control the output waveform. A related unit of equipment is termed a pattern-playback device. It translates a real or a simulated audio spectrogram into an electrical waveform for reproduction by a speaker. A more recent development employs a digital computer for synthesis of artificial speech. This is a major advance from the simple throat vibrator often used to permit intelligible speech by persons who have lost their vocal cords.

APPENDIX I

Resistor Color Codes

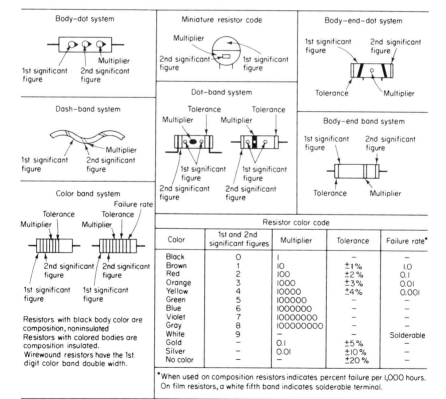

Color	1st and 2nd significant figures	Multiplier	Tolerance	Failure rate*
Black	0	1	−	−
Brown	1	10	±1%	1.0
Red	2	100	±2%	0.1
Orange	3	1000	±3%	0.01
Yellow	4	10000	±4%	0.001
Green	5	100000	−	−
Blue	6	1000000	−	−
Violet	7	10000000	−	−
Gray	8	100000000	−	−
White	9	−	−	Solderable
Gold	−	0.1	±5%	−
Silver	−	0.01	±10%	−
No color	−	−	±20%	−

*When used on composition resistors indicates percent failure per 1,000 hours. On film resistors, a white fifth band indicates solderable terminal.

APPENDIX II

Capacitor Color Codes

Molded mica capacitor codes (capacitance given in MMF)

Color	Digit	Multiplier	Tolerance	Class or characteristic
Black	0	1	20%	A
Brown	1	10	1%	B
Red	2	100	2%	C
Orange	3	1000	3%	D
Yellow	4	10000		E
Green	5		5% (ELA)	F(JAN)
Blue	6			G(JAN)
Violet	7			
Gray	8			I(ELA)
White	9			J(ELA)
Gold		0.1	5% (JAN)	
Silver		0.01	10%	

Class or characteristic denotes specifications of design involving Q factors, temperature coefficients, and production test requirements.
All axial lead mica capacitors have a voltage rating of 300, 500, or 1000 volts, for 4.0 MMF whichever is greater.

Molded paper capacitor codes (capacitance given in MMF)

Molded paper tubular

Color	Digit	Multiplier	Tolerance
Black	0	1	20%
Brown	1	10	
Red	2	100	
Orange	3	1000	
Yellow	4	10000	
Green	5	100000	5%
Blue	6	1000000	
Violet	7		
Gray	8		
White	9		10%
Gold			5%
Silver			10%
No color			20%

1st, 2nd significant figures
Multiplier
Tolerance

1st, 2nd significant voltage figures
Add two zeros to significant voltage figures. One band indicates voltage. Indicates voltage ratings under 1000 volts.

Indicates outer foil. May be on either end. May also be indicated by other methods such as typographical marking or black strip.

Molded-insulated axial lead ceramics

1st, 2nd significant figures
Multiplier
Tolerance
Temperature coefficient

Typographically marked ceramics

Temperature coefficient
Capacity
Tolerance

JAN letter	Tolerance	
	10 MMF or less	Over 10 MMF
C	±0.25 MMF	
D	±0.6 MMF	
F	±1.0 MMF	±1%
G	±2.0 MMF	±2%
J		±5%
K		±10%
M		±20%

Extended range T.C. tubular ceramics

1st, 2nd significant figures
Multiplier
Tolerance
Temp. coeff. multiplier
T.C. significant figure

Color band system

1st, 2nd significant figures
Multiplier
Tolerance

Resistors with black body color are composition, non insulated. Resistors with colored bodies are composition, insulated. Wire-wound resistors have the 1st digit color band double width.

Resistor codes (resistance given in ohms)

Color	Digit	Multiplier	Tolerance
Black	0	1	±2%
Brown	1	10	±1%
Red	2	100	±2%
Orange	3	1000	±3%*
Yellow	4	10000	GMV*
Green	5	100000	±5%
Blue	6	1000000	±8%*
Violet	7	10000000	±12 1/2 %*
Gray	8	0.01 (ELA alternate)	±30%*
White	9	0.1 (ELA alternate)	±10% (ELA alternate)
Gold		0.1 (JAN and ELA preferred)	±5% (JAN and ELA preferred)
Silver		0.01 (JAN and ELA preferred)	±10% (JAN and ELA preferred)
No color			±20%

Extended range T.C. tubular ceramics

Tolerance
1st, 2nd significant figures
Multiplier

Body-end band system

1st, 2nd significant figures
Tolerance   Multiplier

*GMV = guaranteed minimum value, or −0,100% tolerance.
±3, 6, 12 1/2, and 30% are ASA 40,20,10, and 5 step tolerances.

Disc ceramics (5-dot system)	Ceramic capacitor codes (capacity given in MMF)								High capacitance tubular ceramics insulated or non-insulated
	Color	Digit	Multiplier	Tolerance		Temperature coefficient PPM / °C	Extended range Temp. Coeff.		
				10 MMF or less	Over 10 MMF		Significant figure	Multiplier	
1st significant 2nd figures — Multiplier — Tolerance — Temperature coefficient	Black	0	1	±2.0 MMF	±20%	0(NP0)	0.9	−1	
	Brown	1	10	±0.1 MMF	±1%	−33(N033)		−10	
	Red	2	100		±2%	−75(N075)	1.0	−100	
	Orange	3	1000		±2.5%	−150(N150)	1.5	−1000	
	Yellow	4	10000			−220(N220)	2.2	−10000	
Disc ceramics (3-dot system)	Green	5		±0.5 MMF	±5%	−330(N330)	3.3	+1	
	Blue	6				−470(N470)	4.7	+10	
	Violet	7				−750(N750)	7.5	+100	
	Gray	8	0.01	±0.25 MMF		−30(P030)		+1000	
1st significant 2nd figures — Multiplier	White	9	0.1	±1.0 MMF	±10%	General purpose bypass and coupling +100 (P100) (Jan)		+10000	
	Silver								
	Gold								
	Ceramic capacitor voltage ratings are standard 500 volts, far some manufacturers, 1000 volts for other manufacturers, unless otherwise specified.								

High capacitance tubular ceramics insulated or non-insulated

1st significant
2nd figures
Multiplier
Tolerance
Voltage (optional)

Temperature compensating tubular ceramics

1st significant
2nd figures
Multiplier
Tolerance
Temperature coefficient

Current standard JAN and ELA code	Button silver mica	Molded flat paper capacitors (commercial code)	Molded flat paper capacitors (JAN code)
White (ELA) Black(JAN) 1st significant 2nd figures — Multiplier — Tolerance Class or characteristics	1st (when applicable) sig. 2nd for 1st fig. 3rd for 2nd Multiplier Tolerance Class	1st significant 2nd figures — Voltage — Multiplier Black or brown body	Sliver 1st significant 2nd figures — Multiplier — Tolerance Characteristic
Molded ceramics	Button ceramics	Stand-off ceramics	Feed-thru ceramics
Using standard resistor color-code 1st significant 2nd figure — Multiplier —White band Distinguishes capacitor from resistor	1st significant 2nd figures — Multiplier Viewed from soldered surface	1st significant 2nd figures — Multiplier — Tolerance Temperature coefficient	1st significant 2nd figures — Multiplier — Tolerance Temperature coefficient

APPENDIX III

Transistor Identification

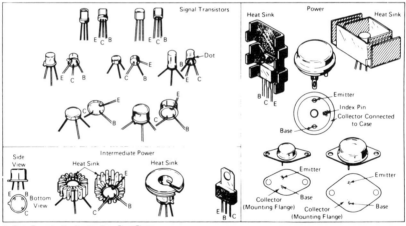

To Test a Transistor You Need to Know Three Things

1. The Basing Configuration (E, B, C, or S, G, D). The Diagram Above Shows Some of the More Common Configurations. If the Transistor Type Number Is Available, the Basing Configuration Can Be Found in the Manufacturer's Handbook. Also, a Schematic May Provide This Information.
2. The Type (NPN or PNP). This Information Can Come From the Circuit, Schematic, or Manufacturer's Handbook.
3. The Power Class. See Diagram Above. (Signal, Intermediate Power, or Power.)

APPENDIX IV

Diode Polarity Identification

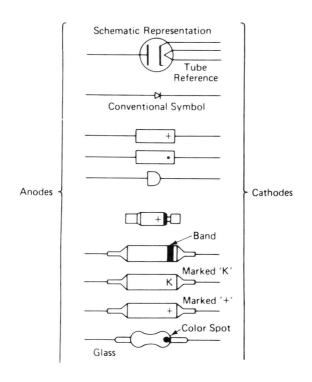

Schematic Representation

Tube Reference

Conventional Symbol

Anodes

Cathodes

Band

Marked 'K'

Marked '+'

Color Spot

Glass

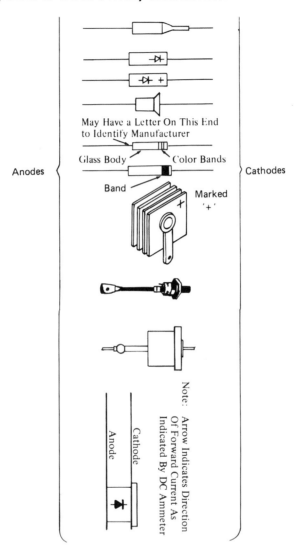

APPENDIX V

Basing Identifications for Typical Transistors

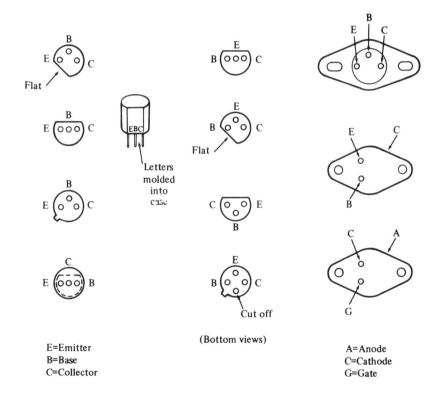

E=Emitter
B=Base
C=Collector

(Bottom views)

A=Anode
C=Cathode
G=Gate

223

Power Ratios, Voltage Ratios, and dB Values

Power Ratio	Voltage Ratio	db − + ← →	Voltage Ratio	Power Ratio
1.000	1.0000	0	1.000	1.000
0.9772	0.9886	.1	1.012	1.023
0.9550	0.9772	.2	1.023	1.047
0.9333	0.9661	.3	1.035	1.072
0.9120	0.9550	.4	1.047	1.096
0.8913	0.9441	.5	1.059	1.122
0.8710	0.9333	.6	1.072	1.148
0.8511	0.9226	.7	1.084	1.175
0.8318	0.9120	.8	1.096	1.202
0.8128	0.9016	.9	1.109	1.230
0.7943	0.8913	1.0	1.122	1.259
0.6310	0.7943	2.0	1.259	1.585
0.5012	0.7079	3.0	1.413	1.995
0.3981	0.6310	4.0	1.585	2.512
0.3162	0.5623	5.0	1.778	3.162
0.2512	0.5012	6.0	1.995	3.981
0.1995	0.4467	7.0	2.239	5.012
0.1585	0.3981	8.0	2.512	6.310
0.1259	0.3548	9.0	2.818	7.943
0.10000	0.3162	10.0	3.162	10.00
0.07943	0.2818	11.0	3.548	12.59
0.06310	0.2512	12.0	3.981	15.85
0.05012	0.2293	13.0	4.467	19.95
0.03981	0.1995	14.0	5.012	25.12
0.03162	0.1778	15.0	5.623	31.62
0.02512	0.1585	16.0	6.310	39.81
0.01995	0.1413	17.0	7.079	50.12
0.01585	0.1259	18.0	7.943	63.10
0.01259	0.1122	19.0	8.913	79.43
0.01000	0.1000	20.0	10.000	100.00
10^{-3}	3.162×10^{-2}	30.0	3.162×10	10^3
10^{-4}	10^{-2}	40.0	10^2	10^4
10^{-5}	3.162×10^{-3}	50.0	3.162×10^2	10^5
10^{-6}	10^{-3}	60.0	10^3	10^6
10^{-7}	3.162×10^{-4}	70.0	3.162×10^3	10^7
10^{-8}	10^{-4}	80.0	10^4	10^8
10^{-9}	3.162×10^{-5}	90.0	3.162×10^4	10^9
10^{-10}	10^{-5}	100.0	10^5	10^{10}

DB Calculations for Power Ratios from 0.01 to 990

Power ratios expressed in + db

Power Ratio	0.0	0.1	0.2	0.3	0.4	0.5	0.6	0.7	0.8	0.9
1	0.000	0.414	0.792	1.139	1.461	1.761	2.041	2.304	2.553	2.788
2	3.010	3.222	3.424	3.617	3.802	3.979	4.150	4.314	4.472	4.624
3	4.771	4.914	5.051	5.185	5.315	5.441	5.563	5.682	5.798	5.911
4	6.021	6.128	6.232	6.335	6.435	6.532	6.628	6.721	6.812	6.902
5	6.990	7.076	7.160	7.243	7.324	7.404	7.482	7.559	7.634	7.709
6	7.782	7.853	7.924	7.993	8.062	8.129	8.195	8.261	8.325	8.388
7	8.451	8.513	8.573	8.633	8.692	8.751	8.808	8.865	8.921	8.976
8	9.031	9.085	9.138	9.191	9.243	9.294	9.345	9.395	9.445	9.494
9	9.542	9.590	9.638	9.685	9.731	9.777	9.823	9.868	9.912	9.956

For power ratios between 0.01 and 0.099, use above table to find db for 100 times the ratio and subtract 20 db.

For power ratios between 0.1 and 0.99, use above table to find db for 10 times the ratio and subtract 10 db.

For power ratios between 1 and 9.9, use above table directly.

For power ratios between 10 and 99, use above table to find db for 1/10th of the ratio and add 10 db.

For power ratios between 100 and 990, use above table to find db for 1/100th of the ratio and add 20 db.

Glossary

"A" signal: The signal fed to the left-hand loudspeaker of a stereo two-speaker system.

ABC: Abbreviation for automatic bass compensation; a circut utilized in some audio equipment to increase the amplitude of the bass tones and make them sound more natural at low volume levels.

Absolute maximum rating: Limiting values of operating conditions, applicable to any device of specified type as defined in its published data.

Absorption: Dissipation of the energy of a sound wave into other forms owing to its reaction with matter.

Absorption coefficient: The fraction of sound power which is absorbed on reflection at any surface. (It ranges between 0 and 1, and unless otherwise stated, is for a frequency of 512 Hz at normal incidence.)

A-B test: Comparison of sound from two sources, such as comparing original program to tape as it is being recorded by switching rapidly back and forth between them.

Acetate backing: A standard plastic base for magnetic recording tape.

Acoustic: Pertaining to sound or to the science of sound.

Acoustic absorption loss: The energy lost by conversion into heat or other forms when sound passes through or is reflected by a medium.

Acoustic absorptivity: The ratio of sound energy absorbed by a surface to the sound energy arriving at the surface. Equal to 1 minus the reflectivity of the surface.

Acoustic attenuation constant: The real part of the acoustic propagation constant; neper per section, or unit distance.

Acoustic capacitance: In a sound medium, a measure of volume displacement per dyne per square centimeter. The unit is centimeter to the fifth power per dyne.

Acoustic clarifier: A system of cones loosely attached to the baffle of a speaker and designed to vibrate and absorb energy during sudden loud sounds, thereby suppressing them.

Acoustic compliance: The measure of volume displacement of a sound medium when subjected to sound waves. Also, that type of acoustic reactance which corresponds to capacitive reactance in an electrical circuit.

Acoustic delay line: A device that retards one or more signal vibrations by causing them to pass through a solid (or liquid).

Acoustic dispersion: The change in speed of sound with frequency.

Acoustic elasticity: The compressibility of the air in a speaker enclosure as the cone moves backwards. Also, the compressibility of any material through which sound is passed.

Acoustic environment: Specification of the significant acoustic properties of a listening area.

Acoustic-electric transducer: A device designed to transform sound energy into electrical energy, and vice versa.

Acoustic feedback: Also called acoustic regeneration. The mechanical coupling of a portion of the sound waves from the output of an audio-amplifying system to a preceding part or input circuit (such as a microphone) in the system. When excessive, acoustic feedback produces a howling sound in the speaker.

Acoustic filter: A sound-absorbing device that selectively suppresses certain audio frequencies while allowing others to pass.

Acoustic frequency response: The voltage-attenuation frequency characteristic measured into a resistive load, producing a bandwidth approaching sufficiently close to the maximum.

Acoustic generator: A transducer such as a speaker, which converts electrical or other forms of energy into sound.

Acoustic horn: Also called a horn. A tube of varying cross section having different terminal areas, which change the acoustic impedance to control the directivity of the sound pattern.

Acoustic impedance: Total opposition of a medium to sound waves. Equal to the force per unit area on the surface of the medium, divided by the flux (volume velocity or linear velocity multiplied by area) through that surface. Expressed in ohms and equal to the mechanical impedance divided by the square of the surface area. One unit of acoustic impedance is equal to a volume velocity of 1 cm ^{3} per s produced by a pressure of 1 μbar. Acoustic impedance contains both acoustic resistance and acoustic reactance.

Acoustic inertance: A type of acoustic reactance that corresponds to inductive reactance in an electrical circuit. (The resistance to movement or reactance offered by the sound medium because of the inertia of the effective mass of the medium.) Measured in acoustic ohms.

Acoustic intensity: The limit approached by the quotient of acoustic power being transmitted at a given time through a given area divided by the area as the area approaches zero.

Acoustic labyrinth: A special speaker enclosure having partitions and passages to prevent cavity resonance and to reinforce bass response.

Acoustic lens: An array of obstacles that refracts sound waves in the same way that an optical lens refracts light waves. The dimensions of these obstacles are small compared to the wavelengths of the sound being focused. Also, a device that produces convergence or divergence of moving sound waves. When used with a speaker enclosure, an acoustic lens widens the beam of the higher-frequency sound waves.

Acoustic line: Mechanical equivalent of an electrical transmission line. Baffles, labyrinths, or resonators are placed at the rear of a speaker enclosure to assist in reproduction of very low audio frequencies.

Acoustic mode: A mode of crystal-lattice vibration that does not produce an oscillating dipole.

Acoustic ohm: The unit of acoustic resistance, reactance, or impedance. One acoustic ohm is present when a sound pressure of 1 dyne per cm^2 produces a volume velocity of 1 cm ^{3} per s.

Acoustic phase constant: The imaginary part of the acoustic propagation constant. The commonly used unit is the radian per second or unit distance.

Acoustic pickup: In nonelectrical phonographs, the method of producing a recording by linking the needle directly to a flexible diaphragm.

Acoustic radiator: In an electroacoustic transducer, the part that initiates the radiation of sound vibration. A speaker cone or an earphone diaphragm are examples.

Acoustic reactance: That part of the acoustic impedance due to the effective mass of the medium, that is, to the inertia and elasticity of the medium through which the sound travels. The imaginary component of acoustic impedance, expressed in acoustic ohms.

Acoustic reflectivity: The ratio of the rate of flow of sound energy reflected from the surface on the side of incidence to the incident rate of flow.

Acoustic refraction: A bending of sound waves when passing obliquely from one medium to another in which the velocity of sound is different.

Acoustic resistance: That component of the acoustic impedance which is responsible for the dissipation of energy due to friction between molecules of the air or other medium through which sound travels. Measured in acoustic ohms; analogous to electrical resistance.

Acoustic resonance: An increase in sound intensity as reflected waves and direct waves combine in phase. May also be due to the natural vibration of air columns or solid bodies at a particular audio frequency.

Acoustic resonator: An enclosure that intensifies those audio frequencies at which the enclosed air is set into natural vibration.

Acoustic scattering: The irregular reflection, refraction, or diffraction of a sound wave in many directions.

Acoustic screen: A plane or curved sheet of sound-absorbent or sound-reflective material used for modification of an acoustic environment.

Acoustic system: Arrangement of components in devices designed to reproduce audio frequencies in a specified manner.

Acoustic transmission system: An assembly of elements adapted to the transmission of sound.

Acoustic treatment: Use of certain sound-absorbing materials to control the amount of reverberation in a room, hall, or other enclosed space.

Acoustic wave: A traveling vibration by which sound is transmitted in air or other medium. The characteristics of these waves may be described in terms of change of pressure, or particle displacement, or of density.

Acoustic wave filter: A device designed to separate sound waves of different frequencies. (Through electroacoustic transducers, such a filter may be associated with electric circuits.)

Acoustics: Science of production, transmission, reception, and effects of sound. Also, in a room or other locations, those characteristics that control reflections of sound waves, and thus the sound reception in the room.

AES: Abbreviation for Audio Engineering Society.

AF: Abbreviation for audio frequency, a range that extends from 20 Hz to 20 kHz.

AFC: Abbreviation for automatic frequency control, a circuit commonly used in FM receivers to compensate for frequency drift to keep the tuner "locked" to a selected station.

Air column: The air space within a horn or an acoustic chamber.

AM: Amplitude modulation; a method of superimposing intelligence on an RF carrier by amplitude variation of the carrier.

Ambience: Acoustic environment.

Ambient noise: Acoustic noise in a room or other location. Usually measured with a sound-level meter.

Ambient sound signal: The difference between the L and R signals.

Amplification: Magnification or enlargement.

Amplifier: An electronic device that magnifies or enlarges audio voltage or power signals.

Amplitude: Also called *peak value:* the maximum value of a waveform (with respect to one polarity).

Anechoic enclosure: A low-reflection audio-frequency enclosure.

Anechoic room: A room in which reflected sound energy is negligible; used for measurement of speaker and microphone characteristics.

Articulation: The percentage of speech units understood by a listener.

Attack: Related to *rise time.* The period of time during which a tone increases to full amplitude after a musical instrument starts to emit a tone.

Attenuation: Opposite of amplification; reduction of audio voltage or power.

Audio: Pertaining to frequencies corresponding to a normally audible sound wave. These frequencies range approximately from 15 Hz to 20 kHz.

Audio level meter: An instrument that measures audio-frequency power with reference to a predetermined level. Usually calibrated in decibels.

Audiophile: One who enjoys experimenting with high-fidelity equipment and who is likely to seek the best possible reproduction.

Audio processor: A specialized type of audio amplifier that eliminates much of the background noise from recordings, and provides volume expansion.

Audio rectification: Spurious demodulation of interfering radio-frequency voltages in an audio system.

Autotransformer: A transformer designed with a single, tapped winding that serves as both primary and secondary.

Audioscope: An oscilloscope built into a stereo or stereo-quad system for monitoring signal amplitude and separation.

Axis (of microphone or loudspeaker): Perpendicular line through center of diaphragm.

"B" signal: The signal fed to the right-hand loudspeaker of a stereo.

Background noise: Noise inherent in any electronic system.

Backloaded horn: A speaker enclosure arrangement in which the sound from the front of the cone feeds directly into the room, while the sound from the rear feeds into the room via a folded horn.

Back loading: A form of horn loading particularly applicable to low-frequency speakers; the rear radiating surface of the speaker feeds the horn and the front part of the speaker is directly exposed to the room.

Baffle: A partition or enclosure in a speaker cabinet that increases the length of the air path from the front to the rear radiating surfaces of the speaker.

Bass: The lower or pedal tones provided by an organ.

Bass boost: A manual adjustment of the amplitude-frequency response of an audio system or component to accentuate the lower audio frequencies.

Bass compensation: An emphasis of the low-frequency response of an audio amplifier at low volume levels to compensate for the lowered sensitivity of the ear to weak low-frequency sounds.

Bass distortion: How clearly a speaker reproduces deep bass tones.

Bass half-loudness points: The low frequency in Hz at which the total sound power of a loudspeaker rolls off to become half as loud as in the rest of the tonal spectrum.

Bass-reflex enclosure: A speaker cabinet enclosure in which a portion of the radiation from the rear of the cone is channeled to reinforce the bass tones.

Bass response: The extent to which a speaker or audio amplifier processes low audio frequencies.

Bassy: A term applied to sound reproduction in which the low-frequency tones are overemphasized.

Beat: A successive rising and falling of a wave envelope due to alternate reinforcements and cancellations of two or more component frequencies.

Bias: An electrical signal of relatively high frequency applied to magnetic tape during the recording process, along with the audio signal. The bias frequency is generally in the range of 70 to 120 kHz.

Bidirectional microphone: One which is live on the front and back, but which is dead at the sides and above and below.

Binaural: A type of sound recording and reproduction. Two microphones, each representing one ear and spaced about 6 in. apart, are used to pick up the sound energy to be recorded on separate tape channels. Playback is accomplished through separate amplifiers (or a two-channel amplifier) or through special headphones wired for binaural listening.

Blocked impedance: The input impedance of a transducer when its output is connected to a load of infinite impedance.

Blocked resistance: Resistance of an audio-frequency transducer when its moving elements are restrained so that they cannot move.

Cabinet, tone: A speaker enclosure designed for operation with an electronic organ.

Capacitor (obs: condenser): Any device designed for storage of electrostatic field energy.

Capstan: The spindle or shaft of a tape transport mechanism that pulls the tape past the heads.

Capture ratio: An FM tuner's ability to reject unwanted co-channel signals. If an undesired signal is more than 2.2 dB lower than a desired signal, the undesired signal will be completely rejected.

Cardioid microphone: Microphone with a semiheart-shaped polar diagram.

Cardioid pattern: A heart-shaped directional pickup pattern for a microphone that assists in reducing background noise.

Cartridge: A transducer device used with a turntable to convert mechanical channels in a disc into electrical impulses.

Cassette: Preloaded container with tape and spools for use on cassette tape recorders. Actually, it is a miniature reel-to-reel tape system.

CCIR characteristics: In tape recording, the preemphasis and subsequent deemphasis standard used in Britain and Continental Europe.

Ceramic: A piezoelectric element that is used as the basis of some phonograph pickups; it generates a potential difference when stressed or strained.

Changer: A record-playing device that automatically accepts and plays up to 10 or 12 discs sequentially.

Channel: A complete sound path. A monophonic system has one channel, a stereophonic system has two, and a quadraphonic system has four. Monophonic material may be played through a stereophonic system, and quadraphonic material may be played through a stereophonic system. An amplifier may have several inputs, such as microphone(s), tuner; mono, stereo, and quad tape; and phono.

Channel balance: Equal response from left and right channels of a stereo amplifier. A balance control in a stereo amplifier permits adjustment for uniform sound volume from both speakers or a hi-fi system.

Chiff: A transient enhancement of certain harmonics in an electronic-organ voice to simulate the timbre of a pipe-organ voice.

Compensator: A fixed or variable circuit built into a preamplifier that compensates for bass and treble alterations that were made during the recording process.

Complex tone: An audio waveform composed of a fundamental frequency and a number of integrally related harmonic frequencies (a pitch and a number of related overtones).

Compliance: Physical freedom from rigidity that permits a stylus to track a record groove precisely, or of a speaker to respond to an audio signal precisely.

Cone: The diaphragm that sets the air in motion to generate a sound wave in a direct-radiator speaker; usually conical in shape.

Conical horn: A horn, the cross section of which increases as the square of its axial length.

Control amplifier: Same as integrated amplifier.

cps: Abbreviation for cycles per second; *see* hertz, cycle, and cycles per second.

CrO_2: Abbreviation for chromium-dioxide magnetic recording tape.

Crossover distortion: Distortion that occurs in a push-pull amplifier at the points of operation where the signals cross over the zero axis.

Crossover frequency: In reference to electrical dividing networks, the audio frequency at which equal power is delivered to each of the channels or speakers.

Crossover network: Filtering circuit that selects and passes certain ranges of audio frequencies to the speakers that are designed for the particular ranges.

Crosstalk: In stereo high-fidelity equipment, crosstalk signifies the amount of left-channel signal that leaks into the right channel, and vice versa.

Crystal: A natural piezoelectric element that is used in some phono pickup cartridges and microphones.

Crystal loudspeaker: A speaker in which piezoelectric action is used to produce mechanical displacement. Also termed a piezoelectric loudspeaker.

Cycle: One complete reversal of an alternating current, including a rise to maximum in one direction, a return to zero, a rise to maximum in the other direction, and another return to zero. The number of cycles occurring in 1 s is defined as the frequency of an alternating current. The word *cycle* is commonly interpreted to mean cycles per second, in which it is a measure of frequency. The preferred term is hertz.

Cycles per second: An absolute unit for measuring the frequency or "pitch" of a sound, various forms of electromagnetic radiation, and alternating electric current. *See* hertz.

Damping: Prevention of vibrations, responses, or resonances that would cause distortion if unchecked. Mechanical control is by friction; electrical control is by resistance.

Damping factor: For any underdamped motion during any complete oscillation, the quotient obtained by dividing the logarithmic decrement by the intensity over the whole sphere surrounding the projector to the intensity on the acoustic axis.

Diffusion (of sound): The extent to which sound waves are broken up by uneven surfaces.

DIN (Deutsche Industrie Normen): German industrial standard.

Direct-radiator speaker: A speaker in which the radiating element acts directly on the air instead of relying on any other element, such as a horn.

Distortion: Deviations from an original sound that occur in the reproduction process. Harmonic distortion disturbs the original relationships between a tone and other tones naturally related to it. Intermodulation distortion introduces new tones that result from the beating of two or more original tones.

Dividing network: Same as *crossover network*.

Dominant Harmonic: A harmonic frequency that has a greater amplitude than the fundamental frequency.

Doppler tone cabinet: A tone-cabinet design in which one or more speakers are rotated or in which a baffle is rotated to produce a mechanical vibrato/tremolo effect.

Doubling: The generation of a large amount of second-harmonic distortion owing to a nonlinear motion of a speaker cone.

Drone cone: An undriven speaker cone mounted in a bass-reflex enclosure.

Dubbing: Copying of previously recorded material. In tape recording, playing a recorded tape on one machine while recording it on another.

Ducted port: A form of bass-reflex speaker enclosure in which a tube is mounted behind the reflex port.

Dynamic cartridge (electrodynamic): A magnetic phono pickup in which a moving coil in a magnetic field generates voltages to form an audio signal.

Dynamic microphone: A microphone that operates on the same basic principle as a dynamic cartridge.

Dynamic range: The range of volumes in a sound passage.

Dynamic speaker: Also termed *a moving-coil speaker*. The moving diaphragm is attached to a coil, which is conductively connected to the source of electric energy and placed in a constant magnetic field. The current through the coil interacts with the magnetic field, causing the coil and diaphragm to move back and forth in accordance with the current variations through the coil.

Dyne per square centimeter: The unit of sound pressure. Originally called a *bar,* but now termed by the full expression.

Echo: A delayed repetition (sometimes several rapid repetitions) of the original sound.

Effective current: The value of alternating or varying current that will produce the same amount of heat as the same value of direct current. Also called *rms current.*

Effective sound pressure: The root mean square of the instantaneous sound pressure at one point over a complete cycle. The unit is the dyne per square centimeter.

Effective value: Also called the rms (root mean square) value. A value of alternating current that will produce the same amount of heat in a resistance as the corresponding value of direct current.

Efficiency: In a speaker, the ratio of power applied to the input terminals expressed as a percentage.

Eigentone (German): The fundamental resonance frequency associated with any dimensional resonance of a room.

Electroacoustic: Pertaining to a device, as a speaker, that involves both electric current and sound-frequency pressures.

Electroacoustic transducer: A device that receives excitation from an electric system and delivers its output to an acoustic system, or vice versa.

Electromagnetic: Pertaining to a phenomenon that involves the interaction of electric and magnetic field energy.

Electrostatic speaker: A type of speaker in which sound is produced by charged plates that are caused to move while one is changed from positive to negative polarity, resulting in forces of attraction or repulsion.

Electrostatic tweeter: A speaker with a movable flat metal diaphragm and a nonmovable metal electrode capable of reproducing high audio frequencies. The diaphragm is driven by the varying high voltage that is applied to the plates.

Enclosure: A housing that is acoustically designed for a speaker or speakers. Also called a tone cabinet in electronic organ technology.

Equalization: Correction for frequency non-linearity of recordings. Phonograph records are cut with low frequencies attenuated and high frequencies boosted. Playback equalization compensates for this, producing a flat frequency characteristic.

Equalizer: An elaborate form of tone control that provides numerous regions of amplitude variation within the audio-frequency range.

Equal-loudness contours: Fletcher-Munson curves, *q.v.*

Erase head: The leadoff head in a tape recorder that erases previous recordings from the passing tape by generating a strong and random magnetic field.

Excess sound pressure: The total instantaneous pressure at a point in a medium containing sound waves, minus the static pressure when no sound waves are present. The unit is the dyne per square centimeter.

Expander (volume): An audio amplifier designed to provide an output level that increases out of proportion to the input level.

Exponential: An exponential curve follows the progress of natural unrestrained growth or decay.

Fast decay: A rapid attenuation of a tone after its keyswitch has been released.

Feed reel: The reel in a tape recorder that supplies the tape.

FET (field-effect transistor): A transistor of the voltage-operated-device classification, instead of the current-operated type as a bipolar transistor.

Fidelity: The faithfulness of sound reproduction.

Filter network: A reactive network that is designed to provide specified attenuation to signals within certain frequency limits; basic filters are termed low-pass, high-pass, bandpass, and band-reject designs.

Flare factor: A number that expresses the degree of outward curvature of speaker horn.

Flat response: A characteristic of an audio system whereby any tone is reproduced without deviation in intensity for any part of the frequency range that it covers.

Fletcher-Munson curves: Also called equal-loudness contours. A group of sensitivity curves showing the characteristics of the human ear for different intensity levels between the threshold of hearing and the threshold of feeling. The reference frequency is 1 kHz.

Flutter: A form of distortion caused when a tape transport or a turntable is subject to rapid speed variation.

FM: Frequency modulation.

FM sensitivity: The minimum input signal required in an FM receiver to produce a specified output signal having a specified signal-to-noise ratio.

FM stereo: Broadcasting over FM frequencies of two sound signals within a single channel. A *multiplexing* technique is utilized.

Folded horn: A type of speaker enclosure that employs a horn-shaped passageway that improves bass response.

Force factor (of an electroacoustic transducer): The complex quotient of the force required to block the mechanical or acoustic system, divided by the corresponding current in the electrical system. The complex quotient of the resultant open-circuit voltage in the electric system divided by the velocity in the mechanical or acoustic system.

Force-summing device: In a transducer, the element directly displaced by the applied stimulus.

Formant: A characteristic acoustic resonance region.

Formant filter: A wave-shaping network or device that changes the waveform of a tone-generator signal into a desired musical tone waveform.

Free impedance: Also called *normal* impedance. The input impedance of a transducer when the load impedance is zero.

Free motional impedance: The complex remainder after the blocked impedance of a transducer has been subtracted from the free impedance.

Free sound field: A field in a medium free from discontinuities or boundaries. In practice, it is a field in which the boundaries cause negligible effects over the region of interest.

Frequency: The number of complete vibrations or cycles completed in 1 s by a waveform, and measured in hertz.

Frequency modulation: A method of broadcasting that varies the frequency of the carrier instead of its amplitude. FM is the selected high-fidelity medium for broadcasting high-quality program material.

Frequency range: The limiting values of a frequency spectrum, such as 20 Hz to 20 kHz.

Frequency response: The frequency range over which an audio device or system will produce or reproduce a signal within a certain tolerance, such as $\pm dB$.

Front-end overload: Distortion or interference caused by an FM tuner's inability to handle strong signals from a nearby transmitter.

Fundamental: The normal pitch of a musical tone; usually, the lowest frequency component of a tonal waveform.

Gain: The value of amplification that a signal obtains in passage through an amplifying stage or system.

Gate circuit: A circuit that operates as a selective switch and permits conduction over a specified interval.

Generator: A tone or signal source, such as an oscillator, frequency divider, or magnetic tone wheel.

Grill: A decorative and protective sound-transparent structure and/or mesh that forms the front surface of a speaker enclosure.

Harmonic: A frequency component of a complex waveform that bears an integral relation to the fundamental frequency. Also called *overtone*.

Harmonic distortion: See distortion.

Head: Electromagnetic device used in magnetic tape recording to convert an audio signal to a magnetic pattern, and vice versa.

Headphones: Small sound reproducers resembling miniature speakers used either singly or in pairs, usually attached to a headband to hold the phones snugly against the ears. Available in monophonic or stereophonic design.

Hertz: A unit of frequency equal to 1 cycle per second (cps).

High fidelity: The characteristic that enables an audio system to reproduce sound as nearly like the original as possible.

High-frequency overload: In a tape recorder, distortion caused by strong high-frequency signals combining to produce undesired new tones. Also called high frequency intermodulation distortion.

Hiss: High frequency noise.

Hole-in-the-middle effect: The lower volume or absence of sound between the left and right speakers of a stereo system.

Horn: Also called *an acoustic horn*. A tubular or rectangular enclosure for radiation of acoustic waves.

Horn cutoff frequency: A frequency below which an exponential horn will not function correctly because it fails to provide for proper expansion of the sound waves.

Horn loading: A method of coupling a speaker diaphragm of the listening space by an expanding air column that has a small throat and a large mouth.

Horn mouth: The wide end of a horn.

Horn speaker: A speaker in which a horn couples the radiating element of the medium.

Horn throat: The narrow end of a horn.

Hum: Noise generated in an audio or other electronic device by a source or sources of electrical disturbance.

Hypercardioid: Polar response of microphone, intermediate between figure-of-eight and true cardioid.

IC: Abbreviation for *integrated circuit*. Integral solid-state units that include transistors, resistors, semiconductor diodes, and often capacitors, all of which are formed simultaneously during fabrication.

IM distortion: Intermodulation distortion. Signals in output caused by interaction of two or more input signals but not harmonically related to them. Expressed as a percentage of the total signal output.

Integrated amplifier: An audio preamplifier and power amplifier housed in a single cabinet.

Integrated receiver: An integrated receiver contains an FM/AM tuner, preamplifier, and power amplifier in the same cabinet.

Intensity (of sound): The sound energy crossing a square meter.

Loudspeaker dividing network: Equivalent term for crossover network.

Loudspeaker impedance: Equivalent term for speaker impedance.

Loudspeaker system: Equivalent term for speaker system.

Louver: The grille of a speaker.

Magnetic armature speaker: A speaker comprising a ferromagnetic armature actuated by magnetic attraction.

Magnetic speaker: A speaker in which acoustic waves are produced by mechanical forces resulting from magnetic reaction.

Magnetic tape: Plastic tape with an iron-oxide coating for magnetic recording.

Manual player: Manual record-playing device used with a changer-type machine.

Mean free path: The average distance that sound waves travel between successive reflections in an enclosure.

Megohm: A multiple unit that denotes 1 million Ω.

Mel: A unit of pitch; a simple 1-kHz tone, 40 dB above a listener's threshold, produces a pitch of 1,000 mels. The pitch of any sound that is judged by the listener to be n times that of a 1-mel tone is denoted as n mels.

Micro: A prefix that denotes one millionth.

Microbar: A unit of pressure commonly used in acoustics. One microbar is equal to 1 dyne per cm.

Microphone: Electro-acoustic transducer.

Milli: A prefix that denotes one thousandth.

Mixing: A blend of two or more electrical signals or acoustic waves.

Modulation: A process wherein low-frequency information is encoded into a higher-frequency carrier or subcarrier; subdivisions include amplitude, frequency, and phase modulation, with various combinations and derivatives thereof.

Monophonic: A recording and reproduction system in which all program material is processed in one channel.

Monorange speaker: A speaker that provides the full spectrum of audio frequencies.

Moving-coil speaker: Also termed a *dynamic speaker*. A speaker in which the moving diaphragm is attached to a coil, which is driven by audio-frequency currents. These currents interact with a fixed magnetic field and cause the diaphragm to vibrate in unison.

MPX: Abbreviation for multiplex.

MRIA: Magnetic Recording Instruments Association.

Multiplexing: A system of broadcasting in which two or more separate channels are transmitted on one FM carrier, as in stereophonic broadcasting.

Multivibrator: A relaxation oscillator, usually developing a semisquare wave form. Subclassifications include the astable, monostable, and bistable types.

Muting: A silencing process or action.

NAB curve: Tape-recording equalization curve established by the National Association of Broadcasters.

NARTB characteristics: In tape recording, the preemphasis and subsequent deemphasis standards used in America and Japan.

Near field: The acoustic radiation field close to the speaker or some other acoustic source.

Neon lamp: A gas diode that emits an orange glow, and operates as an indicator, protective switch, regulator, relaxation oscillator, or divider.

Network: A comparatively elaborate electrical or electronic circuit arrangement.

Noise: Unwanted signal consisting of a mixture of random electrical signals. Also the sum of all unwanted signals such as hum, hiss, rumble, interference, distortion, etc.

Ohm: The unit of electrical resistance, defined as a unitary voltage/current ratio.

Oscillator: An electronic, electrical, or mechanical generator of an electrical signal.

Output: A connection or conductor through which an electrical signal emerges from an electrical or electronic device, circuit, or system.

Overall loudness level: A measure of the response of human hearing to the strength of a sound. It is scaled in phons and is an overall single evaluation calculated for the level of sound pressure of several individual bands.

Overtone: Same as *harmonic*.

Partial: Any one of the various frequencies contained in a complex wave form that corresponds to a musical tone.

Patch cord: A shielded cable utilized to connect one audio device to another.

Peak sound pressure: The maximum absolute value of instantaneous sound pressure for any specified time interval. The most common unit is the microbar.

Permanent-magnet speaker: A moving-conductor speaker in which the steady magnetic field is produced by a permanent magnet.

pF: Abbreviation for *picofarad*.

Phantom channel: An L + R signal formed in a mixer, used to drive a "hole-in-the-middle" speaker.

Phase: Position occupied at any instant in its cycle by a periodic wave; a part of a sound wave or signal with respect to its passage in time. One signal is said to be in phase with, or to lead, or to lag, another reference signal.

Phase inverter: An amplifier that provides an output which is 180 degrees out of phase with its input, or an amplifier that provides a pair of output voltages which are also 180 degrees out of phase with each other.

Phon: The unit for measurement of the apparent loudness level of a sound. Numerically equal to the sound-pressure level, in decibels relative to 0.0002 μbar, of 1-kHz tone that is considered by listeners to be equivalent in loudness to the sound under consideration.

Pickup cartridge: A device used with a turntable to convert mechanical variations into electrical impulses.

Picofarad: A unit equal to 1 micromicrofarad.

Piezoelectric speaker: A speaker that employs a piezoelectric substance as a driver or motor.

Piston action: The movement of a speaker cone or diaphragm when driven at the bass audio frequencies.

Pitch: That characteristic of a sound which places it on a musical scale.

Playback head: The last head of a tape recorder, or the only head on a tape player, which converts the magnetic pattern impressed on a passing tape into an audio signal.

Plug-type connector: A mating connector for a jack.

PM: Permanent magnet.

Polyester backing: A plastic material used as a base for magnetic recording tape.

Port: An opening in the baffle of a bass-reflex speaker enclosure for selective radiation of sound waves.

Power: A unit of the rate at which work is done, or energy is consumed, or energy is generated; electrical power is measured basically in rms watts.

Power amplifier: An amplifier that drives a speaker in an audio system.

Power bandwidth: Specification of a higher-frequency limit and a lower-frequency limit for an amplifier, between which the harmonic distortion is no greater at -3 dB of maximum rated power than the harmonic distortion measured at 1 kHz and at maximum rated power output.

Power output: The signal power delivered by an audio amplifier, measured in watt units.

Power supply: A source of electrical energy; usually, an arrangement that converts alternating current into virtually pure direct current.

Preamplifier: Amplifying arrangement that steps up a very weak input signal to a suitable level for driving an intermediate amplifier or a power amplifier.

Preemphasis: A deliberate exaggeration of the high-frequency components in an audio signal.

Presence: The quality of naturalness in sound reproduction. When the presence of a system is good, the illusion is that the sounds are being produced intimately at the speaker.

Print-through: Magnetization of a layer of tape by an adjacent layer.

Psychophysics: The science of interrelations between physical processes and mental processes.

Pulse: An electrical transient, or a series of repetitive surges.

Quadraphonic: A system whereby sound that is picked up by four separate microphones is recorded on separate channels and played back through separate channels that drive individual speakers.

Quality: Relates to the harmonic content of a complex tonal wave form; also termed *timbre*.

Quarter-track recorder: A tape recorder that utilizes one quarter of the width of the tape for each recording; in stereo operation, two of the four tracks are used simultaneously.

Quieting: Standard of separation between background noise and the program material from a tuner.

Record head: The second head of a tape recorder; used to convert an audio signal to a magnetic pattern on the passing tape.

Recording amplifier: An amplifying section in a tape recorder that prepares an audio signal for application to the record head, and bias current to the erase head.

Record-playback head: A head on a tape recorder that performs both recording and playback functions.

Reference acoustic pressure: That magnitude of a complex sound that produces a sound-level meter reading equal to the reading that results from a sound pressure of 0.0002 dyne per cm at 1 kHz. Also called *reference sound level.*

Relative pitch: The ability to judge one pitch by reference to another. Total lack of this ability is "tone deafness."

Relay: An electromagnetically operated switching device.

Reproducer: A device used to translate electrical signals into sound waves.

Resultant: Denotes a tone that is produced when two notes one fifth apart and an octave higher than the desired note are sounded to produce the desired pitch; a mode of generating *synthetic bass.*

Reverberation: The persistence of sound due to repeated reflections from walls, ceiling, floor, furniture, and occupants in a room.

Reverberation period: The time required for the sound in an enclosure to decay to one millionth (60 dB) of its original intensity.

Reverberation strength: The difference between the level of a plane wave that produces in a nondirectional transducer a response equal to that produced by the reverberation corresponding to a 1-yard range from the effective center of the transducer.

Reverberation time: For a given frequency, the time required for the average sound-energy density, originally in a steady state, to decay to one-millionth (-60 dB) of its initial value after the source is stopped.

RIAA curve: Standard disc-recording curve specified by the Record Industry Association of America.

Ribbon tweeter: A high-frequency speaker, often horn-loaded, in which a stretched, straight, flat ribbon is utilized instead of a conventional voice coil.

Rolloff: The rate at which a frequency response curve decreases in amplitude; usually stated in dB per octave or dB per decade.

Rumble: A low-frequency vibration originating from a vibrating electric motor in a turntable.

Rumble filter: A low-frequency filter circuit designed to minimize or to eliminate rumble interference.

Sabin (square-foot unit of absorption): A measure of the sound absorption of a surface. It is equivalent to 1 sq. ft. of a perfectly absorptive surface.

Scratch filter: A high-frequency filter circuit that minimizes scratchy sounds in playback of deteriorated discs.

Sectoral horn: A horn with two parallel and two diverging sides.

Selectivity: A measure of the ability of an electronic device to select a desired signal and to reject adjacent interfering signals; also termed *bandwidth.*

Sensitivity: The minimum value of input signal that is required by an electronic unit, such as a tuner, to deliver a specified output signal level.

Separation: The degree to which one channel's information is excluded from another channel; customarily expressed in dB units.

Signal-to-noise ratio: The extent to which program material exceeds the background noise level; customarily expressed in dB units.

Sine wave: Graphical representation of simple harmonic motion.

Skating: A sudden displacement of a stylus from one groove to another groove.

Soft-suspension speaker: A speaker design without inherent springiness; it utilizes the reaction of a trapped backwave for restorative force.

Sone: A unit of loudness, a simple 1-kHz tone, 40 dB above a listener's threshold, produces a loudness of 1 sone. The loudness of any sound that is judged by the listener to be n times that of the 1-sone tone is n sones.

Sound: Also called a sound wave. An alteration in pressure, stress, particle displacement, or velocity, propagated in an elastic medium. Also called a sound sensation. The auditory sensation evoked by a sound wave.

Sound absorption: The conversion of sound energy into some other form (usually heat) in passing through a medium or on striking a surface.

Sound absorption coefficient: The incident sound energy absorbed by a surface or a medium, expressed as a fraction.

Sound pressure level: In decibels, 20 times the logarithm of the ratio of the pressure of a sound to the reference pressure, which must be explicitly stated (usually, either 2×10^4 or 1 dyne per cm²). Also, the pressure of an acoustic wave stated in terms of dynes per square centimeter or microbars.

Sound reflection coefficient: Also called *acoustical reflectivity*. Ratio at which the sound energy reflected from a surface flows on the side of incidence to the incident rate of flow.

Sound-reproducing system: A combination of transducers and associated equipment for reproducing prerecorded sound.

Sound spectrum: The frequency components included within the range of audible sound.

Speaker: An electroacoustic transducer that radiates acoustic power into the air.

Speaker efficiency: Ratio of the total useful sound radiated from a speaker at any frequency to the electrical power applied to the voice coil.

Speaker impedance: The rated impedance of the voice coil in a speaker.

Speaker system: A combination of one or more speakers and all associated baffles, horns, and dividing networks used to couple together the driving electric circuit and the acoustic medium.

Speaker voice coil: In a moving-coil speaker, the component that is moved back and forth in response to the applied audio voltage.

Specific acoustic impedance: Also called unit-area acoustic impedance. The complex ratio of sound pressure to particle velocity at a point in a medium.

Specific acoustic reactance: The imaginary component of the specific acoustic impedance.

Specific acoustic resistance: The real component of the specific acoustic impedance.

Speed accuracy: How closely a turntable or tape deck maintains the standard speeds.

Squawker: A midrange speaker.

Standing waves: Reflected waves that alternately cancel and reinforce at various distances.

Stereophonic sound: A system wherein sound energy that is picked up by two separated microphones is recorded on separate channels and is then played back through separate channels that drive individual speakers.

Strength of a simple sound source: The rms magnitude of the total air flow at the surface of a simple source in cubing meters per second, where a simple source is taken to be a spherical source, the radius of which is small compared with one-sixth wavelength.

Strength of a sound source: The maximum instantaneous rate of volume displacement produced by the source when emitting a sinusoidal wave.

Stroboscopic disc: A cardboard or plastic disc with a specialized printed design suitable for checking turntable speed.

Stylus: Same as phonograph needle.

Subharmonic: An integral submultiple of the fundamental frequency in a tonal waveform.

Subordinate fundamental: A fundamental frequency that has less amplitude than one or more of the associated harmonics.

Supertweeter: A speaker designed to reproduce the highest frequencies in the audio range.

Surround sound: Same as stereophonic sound.

T-pad: A three-element fixed attenuator.

Take-up reel: A reel on a tape recorder that winds the tape after it passes the heads.

Tape deck: A tape unit without a power supply or speaker.

Tape recorder: A tape machine that provides both recording and playback facilities.

Tent: A group of acoustic screens, arranged to trap (and usually to absorb) sound in the region of the microphone.

Terminal: Electrical connection point.

Timbre: Also termed *tone color*. The distinguishing quality of a sound that depends primarily upon harmonic content and secondarily upon volume.

Tone: The fundamental frequency or pitch of a musical note.

Tone arm: A pivoted arm on a turntable that houses the pickup cartridge.

Tone burst: A test signal comprising short sequences of sine-wave energy.

Tone color: Also termed *timbre*. Classified as *diapason, flute, string,* or *reed*.

Tone control: A control that provides variation of an amplifier's frequency response.

Tone generator: An organ section that generates the basic voice wave forms.

Tracking: The path of a phono stylus within the grooves of a disc.

Transducer: A device that converts one form of energy into another form.

Transient: An electrical surge.

Transient response: The ability of a speaker to follow sudden changes in signal level.

Triaxial speaker: A dynamic speaker unit consisting of three independently driven units combined into a single speaker.

Turntable: Same as *record player*.

Tweeter: A speaker designed to reproduce the higher audio frequencies, usually those above 3,000 Hz.

Varistor: A voltage-dependent resistor.

Volume: Same as *expression*. A relative sound level.

Volume expansion: Reproduction of sound with an output level that is disproportionate to its input level. (A phono preamplifier provides volume expansion).

Watt: A power unit, equal to the product of one volt and one ampere.

Wave (sound wave): A succession of compressions and rarefactions propagated through a medium at a constant velocity.

Woofer: A speaker designed to reproduce bass tones.

Wow: A form of distortion that occurs when a magnetic tape varies back and forth in speed, or a turntable varies similarly in rpm.

Wow-wow: A very slow vibrato effect.

Zero beat: Precise equality of two frequencies in a mixture of waveforms (basically a mixture of two sine waveforms).

Index